NUMERICAL METHODS
with VBA Programming

James W. Hiestand
The University of Tennessee at Chattanooga

JONES AND BARTLETT PUBLISHERS

Sudbury, Massachusetts

BOSTON   TORONTO   LONDON   SINGAPORE

World Headquarters
Jones and Bartlett Publishers
40 Tall Pine Drive
Sudbury, MA 01776
978-443-5000
info@jbpub.com
www.jbpub.com

Jones and Bartlett Publishers Canada
6339 Ormindale Way
Mississauga, Ontario L5V 1J2
Canada

Jones and Bartlett Publishers International
Barb House, Barb Mews
London W6 7PA
United Kingdom

Jones and Bartlett's books and products are available through most bookstores and online booksellers. To contact Jones and Bartlett Publishers directly, call 800-832-0034, fax 978-443-8000, or visit our website www.jbpub.com.

Substantial discounts on bulk quantities of Jones and Bartlett's publications are available to corporations, professional associations, and other qualified organizations. For details and specific discount information, contact the special sales department at Jones and Bartlett via the above contact information or send an email to specialsales@jbpub.com.

Production Credits
Acquisitions Editor: Timothy Anderson
Editorial Assistant: Melissa Potter
Production Director: Amy Rose
Production Assistant: Ashlee Hazeltine
Senior Marketing Manager: Andrea DeFronzo
V.P., Manufacturing and Inventory Control: Therese Connell
Composition: Pine Tree Composition, Inc.
Cover Design: Kristin E. Parker
Cover Image: © Boris Djuranovic/ShutterStock, Inc.
Printing and Binding: Malloy Incorporated
Cover Printing: Malloy Incorporated

Library of Congress Cataloging-in-Publication Data
Hiestand, James.
Numerical methods with VBA programming / James Hiestand.
   p. cm.
Includes bibliographical references and index.
ISBN-13: 978-0-7637-4964-4 (pbk.)
ISBN-10: 0-7637-4964-8 (pbk.)
1. Microsoft Visual Basic for applications. 2. Application software—Development. 3. Numerical analysis—Data processing. I. Title.
QA76.76.D47H52 2008
005.2'762—dc22
                2008013281

6048
Printed in the United States of America
12 11 10 09 08 10 9 8 7 6 5 4 3 2 1

Dedication

To our grandchildren

Exact Uncertainty

$\bar{Q}$ = average Q

    ＊ Find Q_{max} & Q_{min}

Absolute = $\Delta Q = \dfrac{Q_{max} - Q_{min}}{2}$

$Q = \bar{Q} \pm \Delta Q$

Relative % = $\dfrac{\Delta Q}{Q} \ast 100$

Approx. Uncertainty

$\dfrac{dQ}{Q}$ = Relative %

Simpson's 3/8 Rule:

N MUST BE 3

$$I = (b-a) * [f(a) + 3*f(at\,\Delta x) + 3*f(at\,2\Delta x) + f($$

$$* E_{TRUN} = -\frac{(b-a)^5 * f^{IV}(x)}{6480}$$

$$f^{IV}_{avg} = \frac{(f'''(b) - f'''(a))}{(b-a)}$$

Richardson Extrapolation & Romberg Intergration

Λ	Δt	k=1	K=2 OR Richardson
1	4	$I_{1,1}$ 80.4719	$I_{1,2}$ 53.19066
2	2	$I_{2,1}$ 60.01097	53.27473 $-I_{1,3}$
4	1	$I_{3,1}$ 54.95879	53.28160
8	0.5	$I_{4,1}$ 53.70091	

FOR k=2 → Richardson Extrapolation

$$I = \frac{4*[60.01097 - 80.4719]}{3} \quad OR \quad \frac{4*[I_{2,1}-I_{1,1}]}{3}$$

for k=3 → Romberg Intergration

$$I = \frac{16*[I_{2,2} - I_{1,2}]}{15}$$

for k=4

$$I = \frac{64*[I_{2,3} - I_{1,3}]}{63}$$

$$\Delta x = \frac{(b-a)}{n}$$

"The purpose of computing is insight, not numbers."

R. W. Hamming

pg 82-83

Trap Rule: $\underline{I = \frac{\Delta x}{2} * Sum}$

$Sum = [f(a) + 2*(fa + \Delta x) + 2*(f(a) + 2\Delta x)...+ f(b)]$

* To find True Error find difference of Integral -"I"

* $E_{TRUN} = \frac{-\Delta x^2 * (b-a) * f''(x)}{12}$

* $E_{TRUN, avg} = -\Delta x^2 * \frac{f'(b) - f'(a)}{12}$

pg 88-89

Simpson's 1/3 Rule: $\frac{\Delta x}{3} * Sum = I$

<u>MUST BE EVEN</u>

$Sum = [f(a) + 4*f(a+\Delta x) + 2*f(a+2\Delta x)...f(b)]$

* $E_{TRUN} = \frac{-\Delta x^4 * (b-a) * f^{IV}(x)}{180}$

$f^{IV}_{avg} = \frac{f'''(b) - f'''(a)}{b-a}$

Contents

Preface

■ Introduction

Numerical Methods with VBA Programming was written to restore the combination of numerical methods and computer programming that is so rarely seen in today's texts. Numerous books are available today that introduce either a high-level programming language or numerical methods. Despite the fact that the two are natural allies for students of engineering and computer science, very few of today's texts combine both. Thankfully, I was introduced to this combination more than 40 years ago when I took an undergraduate course using the text *Numerical Methods with FORTRAN Programming* by Daniel McCracken and William Dorn.

I 'crunched' numbers in industry for twelve years, and for the past twenty-two years I have taught programming and numerical methods at the freshman and sophomore levels. This text is based on that experience. Programming was originally created in the FORTRAN Programming Language, but has since been replaced by C/C++ and VBA, in conjunction with Excel. While the two areas were first taught separately, an hour's reduction in the program has required that the programming and numerical methods be combined. Unfortunately, this requires a reduction in depth and I have yet to find a suitable book that covers both. Therefore, in my classes I have relied on books solely as references and distributed my own notes.

Due to my experience, I am familiar with those areas in numerical methods and programming that prove troublesome to many students, and they will be covered with care. I point out potential programming pitfalls and provide tips for overcoming these problems. In presenting each numerical method, I use simple examples that are transparent enough to clearly demonstrate the method presented. Typically, more than one method for solving the same problem is presented, along with brief comments on the advantages of each solution. There are no 'one-size-fits-all' techniques in numerical methods, though clearly some methods (e.g., Newton–Raphson for non-linear equations) dominate in some areas.

Recalling my own days as a student, I believe that new users of numerical techniques fail to appreciate the importance of error estimation. Perhaps this reflects in the success of the techniques used in the problems presented. For this reason, closed–form solutions are presented whenever possible, allowing students to compare the numerical solutions with the analytic ones. In addition, error estimates for particular methods are provided when available. This allows students to develop confidence in the numerical techniques while providing some insight into the practical accuracy of the methods being considered. Throughout the text closed–form answers are referred to as correct, rather than exact, thereby allowing answers with a few decimal places.

■ Approach and Organization

The language used throughout the text is Visual Basic Application (VBA). The programming examples developed and the problems posed are drawn from the areas of numerical methods. This provides a natural application of the programming features being introduced, while relieving the student from the tedium of performing otherwise long calculations by hand. This interaction should make the study of both areas more productive and pleasant. Relatively short manual exercises are included at the end of Chapters 2–12 to reinforce understanding of the methods presented. Computer problems are also included where relevant.

In this text, VBA is accessed through Microsoft Excel, which is readily available on most computers. Excel and VBA provide an excellent, mid-level general computational tool for both students and professionals, which is discussed throughout the text. For example, results generated in a VBA program may be stored in an Excel spreadsheet to which the chart-making capabilities of Excel may then be applied.

The coverage of VBA in the text is introductory. The basic elements of programming are covered, including: input/output, variable types, arithmetic, decisions, loops, functions and subprograms, and arrays. These are treated in enough detail to allow the application of VBA to the numerical techniques covered. These elements are common to other higher-level computer languages, though the implementation is different. Thus, learning to use them in VBA will enable the student to learn other languages relatively easily. In this sense, the first programming language may be the hardest. Features like forms, which are not common to other higher-level languages, have not been included.

While numerical solutions are always a fallback option, closed–form solutions should be obtained whenever possible—I cannot stress this enough. However, in practical applications it is very common to encounter an integral, or an algebraic or differential equation that cannot be solved in closed–form. Fortunately, when properly understood, numerical techniques are accurate, reliable, and easy to apply.

Numerical techniques rely on fundamental properties of the particular process involved. Therefore, numerical integration techniques exploit the fact that an integral is the area under a curve and seek to evaluate that area accurately. These relationships are stressed as a motivation and as an aid to understanding the methods. Where appropriate, I have included analogies to ordinary, non-technical experiences. For example, arrays can be thought of as members of a family with a common last name. Solving a two-point boundary value problem is much like trying to decide when to leave the house in order to catch a flight departing a certain airport at a given time.

Numerical Methods with VBA Programming is limited to basic techniques suitable for a sophomore engineering or physical science major. Although skills in solving differential equations are not required, a year of calculus and a very basic understanding of differential equations are recommended. The techniques presented are standard and only a few prerequisites are assumed, including: a rearrangement routine for linear equations to be solved using Gaussian–Seidel and an approximate technique I developed for dealing with stiff ordinary differential equations. Estimating a starting value prior to finding the root of a non-linear equation is addressed specifically—an important practical topic rarely covered in other texts. Instead, examples elsewhere typically suggest a starting value seemingly plucked out of the air and then proceed to the solution. Also, the examples of the fourth order Runge–Kutta method in this text include more details than many other texts.

Flowcharts for some simple programs, as well as for several of the numerical procedures discussed, are included in the relevant programming sections. Rather than pseudocode, I've chosen flowcharts to provide general guidance to the programmer without giving away all the details. Flowcharts are also more generic and, therefore, may be used with other programming languages. I have provided a flowchart for my linear equation rearrangement program, which seeks to achieve diagonal dominance prior to applying the Gaussian–Seidel method. I've also included my flowchart for a secant rule program that may be used to solve two simultaneous non-linear equations. Finally, a program to apply the Cash–Karp Runge–Kutta method with variable step-size to two ordinary differential equations is included.

An alternative to the programming of numerical methods is the use of a numerical package such as MATLAB® or Maple™. These packages are powerful and convenient and many texts take this approach. However, to use a 'real-world' example, I prefer to know that when I get on an airplane, the pilots know how to expertly fly the plane and not merely turn on the autopilot—as useful as it may be. I strongly believe that teaching the students to write their own programs to solve numerical problems provides them with a greater understanding of methods and better prepares them for future work where they may need to write specialized software. Additionally, programming requires the careful, logical, step-by-step approach generally needed to solve problems in engineering and computer science, and reinforces this skill.

This text includes more material than I can cover in my one-semester course. Chapter 11 is often rushed and Chapter 12, an introduction to partial differential equations, is normally excluded. I've included the chapters in the text, both for completeness and to provide more examples of the power of computing and arrays when solving a very important group of problems in engineering and computer science. Naturally, individual instructors will select from the topics included and perhaps add some of their own.

Beyond Chapter 3, the chapters are independent of one another and some may be skipped or covered in a different order. The arrangement of material is dictated by the development of the aspects of programming. Linear equations (Chapter 9) are treated late in the text, just before computer arrays are introduced in Chapter 10, where they are used in linear equations solutions. It is my experience that for many students arrays are the most difficult topic in elementary programming.

The appendices conveniently gather together material that is widely available, yet typically isolated from one another. These include my summaries of Excel and VBA commands, Excel functions accessible in VBA, numerical methods with the Casio fx-115MS calculator (allowed in the Fundamentals of Engineering (FE) exam), and differentiation fundamentals.

I sincerely hope that both instructors and students find this text to be a clear, an accurate, and an interesting introduction to these subject areas. I welcome feedback, so please contact me with any questions or concerns.

■ Supplements

Solutions to the Exercises and a PowerPoint Slide Presentation are provided online for instructors at http://www.jbpub.com/catalog/9780763749644.

■ Acknowledgments

I would like to thank everyone who assisted me with this project, including those professors who reviewed the manuscript: Mahesh Aggarwal, Gannon University; and Justin Wozniak, University of Notre Dame.

I'd also like to thank the editorial and production staff at Jones and Bartlett Publishers, especially: Tim Anderson, Acquisitions Editor; Melissa Potter, Editorial Assistant; Amy Rose, Production Director; Melissa Elmore, Associate Production Editor; and Ashlee Hazeltine, Production Assistant.

■ A Note to Instructors

Individual instructors have their own preferences for the form of input, output, and documentation for computer programs. Therefore the exercises in the programming chapters state only the problem to be solved. It is expected that each instructor will provide more detailed requirements for the problems assigned.

1 · Introduction

■ 1.1 Why Numerical Methods?

Engineers and scientists frequently encounter linear and nonlinear mathematical equations, integrals, differential equations, and data to be manipulated. Sometimes the manipulations to be performed are easy and straightforward; often they are not. This is particularly true of nonlinear problems, that is, problems in which variables occur as products, including products of themselves, or as functions of transcendental functions, like the trigonometric or logarithmic relations. If the mathematics to be performed cannot be done in closed form (i.e., an exact analytic symbolic solution is obtained), recourse must be made to numerical approximations.

Fortunately, these methods, when properly understood and used, are powerful and accurate. It should be understood that recourse to a numerical solution is always a fallback position. Analytic solutions are to be preferred because they are more accurate, and once the algebraic expression has been obtained, the solution can be applied to a variety of system parameters. With a numerical solution, changing an input variable means the solution must be started over from the beginning. Crudely, it might be likened to the difference between knowing how to spell a word compared to looking it up in the dictionary each time you needed it.

Integration is an area where recourse must frequently be made to numerical methods. The integral

$$\frac{2}{\sqrt{\pi}} \int_0^z e^{-x^2}\, dx$$

cannot be evaluated analytically, but its value for a particular value of z is needed in probability calculations. We shall look at methods of evaluating such integrals numerically in Chapter 5.

We shall also look at methods of solving nonlinear algebraic equations that cannot be solved in closed form. Often equations of this type involve trigonometric or logarithmic terms. A typical equation of this type is

$$f(x) = \tan(x) * \sin(x) - 2.15, \text{ with } x \text{ in radians.}$$

Several methods of solving such equations will be considered in Chapter 7.

Differential equations are another class of problems for which recourse to numerical methods must be made frequently. An example of a deceptively simple differential equation that requires a numerical solution is the differential equation describing the motion of a simple pendulum. With the angular position denoted by θ and the acceleration due to gravity by g, this equation is

$$\frac{d^2\theta}{dt^2} = -g\sin(\theta).$$

This equation is nonlinear and cannot be solved in closed form though an approximate solution can be obtained for small values of θ by replacing $\sin(\theta)$ by θ to yield $\frac{d^2\theta}{dt^2} = -g * \theta$, which has the simple solution $\theta = \theta_o * \cos\left(\sqrt{g} * t\right)$ where θ_o is the initial angle. We shall examine methods of solving ordinary differential equations in Chapter 8.

Students learn to solve sets of linear equations in algebra classes. A typical set of two linear equations can be represented symbolically as

$$a_{11} * x_1 + a_{12} * x_2 = c_1 \quad \text{and}$$
$$a_{21} * x_1 + a_{22} * x_2 = c_2.$$

But those sets usually included only two or three equations. If the equation set becomes large to include hundreds, thousands, or even millions of equations, clearly we need a systematic method that can be programmed into a computer to solve the equations in the set. We shall consider some of these methods in Chapter 10.

Curve fitting, in which we attempt to find the equation that most closely passes through a series of measured data points, will be investigated in Chapter 11. The techniques there will enable us to approximate the data with linear, polynomial, power law, or exponential equations. A typical problem is shown in Figure 1.1 where the specific heat of water is shown as a function of temperature.

In addition, we shall briefly consider the subject of *splines,* in which the data points are connected by a series of curves of assumed shape. Splines are more likely to be used for drawing smooth curves than as an analysis tool.

Finally, Chapter 12 introduces the solution of *partial differential equations.* This is a big, complex topic, where much current research effort is being made because of its importance in solving numerous difficult problems. Numerical weather forecasting, for example, involves the solution of many partial differential equations.

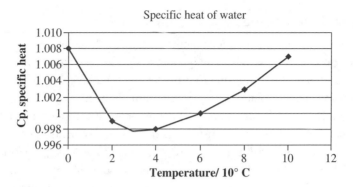

FIGURE 1.1 Curve Fit of the Specific Heat of Water, Cp, as a Function of Temperature.

■ 1.2 Why Programming?

There is an abundance of quality software available to the engineer or scientist today that may be applied to many problems of interest. But, frequently, problems arise that require specialized solutions not available in commercial software or, if available, not conveniently solved. When this occurs, the ability to write one's own program is very convenient. This may be especially true of more advanced problem-solving.

Furthermore, writing a computer program that performs correctly and efficiently requires logical thinking at every step. Therefore, learning to program in virtually any language and then using that ability to write programs to solve problems is a wonderful exercise in the step-by-step logical thinking that is required to solve problems in science and engineering. Writing computer programs is an intellectual workout. Besides, it is fun!

Higher-level computer languages all have certain capabilities in common. We must be able to

perform calculations,

input and output data and results,

make decisions,

perform repetitive calculations in loops,

transfer control to different parts of the program, and

store variables with a common name but distinguished by one or more subscripts.

Each of these topics will be considered using the language chosen for this text, Visual Basic for Applications (VBA). But the general approach to these tasks will apply to any of the higher-level languages.

■ 1.3 Why Numerical Methods and Programming?

Obviously, when one writes a computer program, the solution to some task, whether simple or complex, is sought. Though aimless exercises could be proposed and solved, the application of programming to problems within one's technical field will be more interesting and the solutions potentially useful. Numerical analysis is concerned with manipulating numbers, frequently lots of numbers in complex calculations. Solving these by hand is tedious, time-consuming, and subject to errors. The computer—a wonderful servant if we will just tell it precisely what to do—will perform these calculations both very quickly and very accurately. The combination of the need to manipulate numbers, and the need for programming exercises that are both interesting and useful, makes the combination of learning to program and the learning of numerical analysis a natural and mutually supportive combination.

■ 1.4 Why VBA?

A number of higher-level programming languages are available to engineers and scientists. These include Fortran, C, C++, Java, and Visual Basic, as well as VBA. All these languages are widely used in technical computing, and commercial technical software will certainly be written in one of them.

But VBA has two distinct advantages. It is widely available, being accessible through Microsoft Excel®, which is standard spreadsheet software available on Windows® computers. Furthermore, the general capability of Excel, including its many functions and graphing features, may be combined with the advantages of special-purpose software written by the user to form a powerful and very useful midlevel package for general engineering and scientific analysis. Though VBA is the least powerful of the languages mentioned, its widespread availability; good, user-friendly error checking; and its ability to work with Excel, make it an excellent candidate for a first computer language. Since all higher-level languages have much in common, the first language is the hardest to learn. Once students become competent in using VBA, moving on to other languages as needed should be relatively easy.

VBA is a descendant of the original Beginner's All-purpose Symbolic Instruction Code (BASIC) developed by John G. Kemeny and Thomas E. Kurtz for use at Dartmouth College in 1964. That language was intended to be user-friendly and widely applicable. It also had graphing capability. Present-day VBA is even friendlier, a development made possible by being available through a computer monitor and offering immediate error-checking and color-coded features.

2 Programming Basics: Arithmetic, Input/Output, and All That

■ 2.1 General Remarks

A computer program is a set of instructions issued to a computer to carry out a particular sequence of operations. Programs may be written in many languages. VBA, for example, is easy to use and its availability makes it an excellent "first language" to learn for technical computing. A visual version of the original BASIC, VBA is, in many ways, similar to Fortran, a language widely used in technical computing. VBA is also easily accessed from Microsoft Excel, which is available on most computers using the Windows operating system.

VBA is a higher-level computer language. Higher-level languages aimed at technical computing are written in a form that resembles the algebra that would be used if the problem were to be solved by hand. A set of instructions make up the computer program. No imprecision is allowed, unlike that in most communication among human beings. The vocabulary and syntax must be precisely followed if the program is to be understood by the computer and executed properly. Those instructions must be translated into a language the computer understands. In this sense, it is like an interpreter who translates one human language into another if two humans who do not speak a common language wish to communicate.

The vocabulary of higher-level computer languages like VBA is small. Writing a program (i.e., communicating with the computer) is like talking to a young child or an animal. The vocabulary must be one that the child or animal understands. To a pet, "food is in your dish," may be understood but "meat is in your bowl," may not be; same idea but different words. Fortunately, the number of commands used in a computer language is small. Unfortunately, commands may differ from one computer language to another, just as words do among human languages.

Writing an accurate computer program requires considerable care. Whereas a misplaced comma or misspelled word in written English probably would not change the meaning of a piece of writing very much (except perhaps in a legal document), computer programs must be precisely written, using the correct syntax and vocabulary of the computer language employed. It is like giving instructions to someone on how to go someplace. The instructions must be precise (turning at the second light probably takes you to a different place from turning at the third light). Instructions also must be

5

in a language understood by the listener or, for a program, by the computer. Because both grammatical correctness and precision must be satisfied for the program to work correctly, it takes time to write a program that runs correctly. If a computation needs to be carried out only once, it is probably quicker to do the calculation manually. On the other hand, if the calculation is to be performed frequently with different numbers, it is more efficient to write a program to carry out the sequence of instructions.

Students sometimes become frustrated in writing programs because a computer always does exactly what it is told, (unlike my children when they were at home). If the program fails, it is not enough to say, "But I meant. . . ."

Writing a good program should begin with a plan for the program, just as a paper written for an English or history class should start with an outline. In programming, such an outline may be in the form of a flowchart (applicable to almost any computer language) or in pseudocode, which is a shorthand form of the program similar to the language being used. A good flowchart or pseudocode should include all the logical steps of the program, highlighting especially the branches and decisions that must be made. The components of a flowchart and two examples will be considered, beginning in Section 2.9.

While the program is being written, it may not work properly as a result of mistakes in either language (syntax or spelling violations), logic (faulty instruction sequence), or both. Error messages from the compiler or interpreter will assist in finding the first type of error, usually by identifying the line where the error was detected. Just as a single wrong turn may get you hopelessly lost, some errors will sufficiently confuse the compiler such that a single mistake could generate more than one error message in some programming languages. This can be discouraging, but on the other hand it is quite satisfying to correct a single mistake and find the number of error messages reduced significantly. But when VBA is run, it proceeds through a program one error at a time, because program lines are executed sequentially.

Once the grammatical errors have been eliminated, the program may be executed. It is at this point that any errors in logic become apparent. These tend to be harder to find than those in the computer language grammar. They may be subtle and the computer cannot help, for it does not know what you intended. For the purposes of verifying that a program is working correctly, it should be run with a set of data ("check data") for which the results are known.

If results are different or if the program stops running before it is finished (for example, a division by zero occurs), debugging (eliminating "bugs" or programming errors) must begin. This can be tedious. It is best accomplished by comparing computer-generated results generated at various stages within the program to results achieved by hand at the same point in the program sequence. VBA has some handy debugging tools that allow you to follow the progress of your program. These will be discussed in Section 2.13. In longer programs, intermediate results may be obtained by including extra output from the program that displays results, at critical points in the program on the screen, or in an output file. It is worth labeling these results, for example, "A = 6" instead of merely "6". By working through the program to ensure that computed and hand calculated results agree at each step, the program can be successfully debugged. The check data, the hand calculations, and the computed results that agree should be retained for documenting that the program is running correctly.

In addition to generating check data and results, you should make sure that a computer program includes documentation. This may be elaborate in the form of a user's manual for complex, commercial grade software, or it may be simple, consisting of a few lines of "comment" that briefly tell what the program does, the input required, the output provided, the units used (e.g., degrees Fahrenheit or Celsius), the author (programmer), and the date of the last revision. Though programmers tend to put off writing documentation, it is very important for running a program after it has first been written. It is surprising how quickly we forget, even for programs we have written ourselves, what the input variables are and what their units should be. Brief documentation is included in the sample programs in this text.

For short programs, such as those written in this course, it is convenient to place documentation at the top of the program using comment lines. This has the advantage of the documentation being with the program rather than in a separate file or on paper that may get separated and lost. Of course, for major pieces of software, documentation typically requires the preparation of a proper user's manual.

■ 2.2 Parts of a Computer Program

Regardless of language, computer programs for engineering applications may include the following features. The chapters where these are discussed are also indicated.

Input/Output: Information is given to the computer via keyboard, from a spreadsheet (VBA), and/or input file; results from the computer are output to the screen, spreadsheet (VBA), and/or an output file. Any combination of these methods may be used. (See Chapter 2.)

Arithmetic: The arithmetic operations of addition (+), subtraction (–), multiplication (*), division (/), and raising a number to a power (^). (See Chapter 2.)

Variable types: Computer languages frequently distinguish between integers and numbers containing decimals. The latter may be further divided according to the number of decimal places retained. In addition, programming languages generally accommodate alphanumeric variables (combinations of letters and number, e.g., agent007) and text variables called strings, such as dog, velocity, and force. (See Chapter 2.)

Decisions: Computer programs frequently proceed in different directions depending on the values of the numbers that have been calculated. For example, a positive result may lead to different subsequent steps from a negative result. Computers make decisions by comparing the size of numbers (and the numeric value of letters). (See Chapter 4.)

Loops: Computers are especially good at repetitive calculations. They perform these very accurately (for a correctly written program), very quickly, and they do not become bored. Hence, high-level computer languages offer convenient ways for performing these calculations via loops. Loops are of two general types: those for which we know how many times we want to perform the loop calculations and those for which the number of repetitions depends on the results. By way of analogy: you may decide to run a fixed number of laps around a track, or you work on your checkbook until it is balanced. (See Chapter 4.)

Subprograms and Functions: These are pieces of code that perform specialized tasks and may be reached from other parts of the program. They are the "special teams of programming" called on when needed, just as the punter enters a football game as needed. They may be compiled separately, and once written, they may be used in different programs. They shorten the length of a program by allowing a certain sequence of steps to be written only once, even if needed multiple times at different parts of the program. (See Chapter 6.)

Arrays: Arrays provide a way of using subscripted variables (e.g., x_1, x_2, etc., within a program). They have a common name, with the individual variables being distinguished by the subscript form of the particular language. Multiple subscripts are also allowed, as in a matrix (e.g., $a_{1,2}$). Combined with loops, arrays can significantly simplify a program that must deal with many data of the same nature. (See Chapter 10.)

Each of these program features will be introduced in subsequent chapters, along with examples from numerical analysis to demonstrate their use.

■ 2.3 Opening VBA

You can create a VBA program by selecting "Macros" from the Tools/Macro pull-down menu on the Excel Toolbar. When the gray dialog box (Figure 2.1) displays, type the name of the program—for example, *Prog1*—in the space provided under "Macro name": and click the "Create" tab to select it.

The VBA program sheet will then display as shown in Figure 2.2.

A VBA program is composed of one or more subprograms. The one named in the Create step is the main subprogram. The name of the program selected follows the word *Sub* (which is short for *subprogram*; this program is actually a subprogram). The words *End Sub* are included automatically after a blank space where you will type the program statements. Program statements must be inserted between the *Sub* and *End Sub* lines.

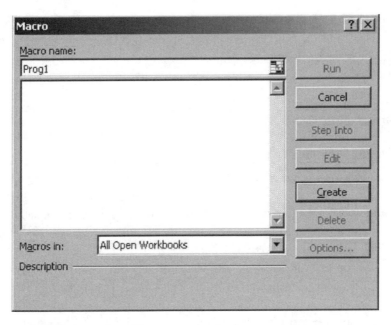

┃ FIGURE 2.1 Macro Control Box.

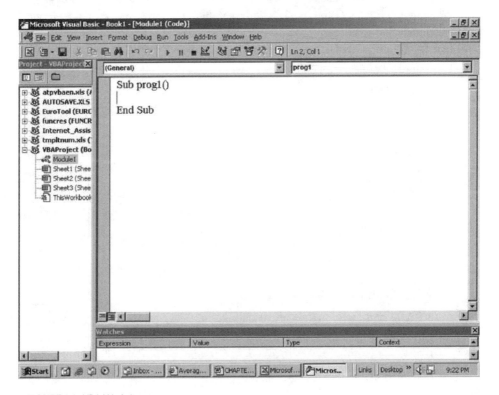

┃ FIGURE 2.2 VBA Worksheet.

If additional subprograms are to be included, you must list them separately, generally below the first *End Sub* line. These will be considered in Chapter 6. Subprograms and functions must not be intertwined.

■ 2.4 VBA Statements

2.4.1 Variable Names

VBA variables must begin with a letter and may be up to 255 characters long. (But do not use long names. Those take time and encourage typing errors.) No breaks or spaces are permitted. After the first character letters, numbers, _ (underscore), and % may be used. If spaces are desired in a variable name, they should be filled with an underscore to make the name continuous, for example, *speed_of_light*. Names should remind the programmer or user of the program what the variable means. *Velocity*, *v6*, and *speed_of_light* are legal names but *6v* (starts with a number) and *speed of light* (spaces) are not.

VBA is not case-sensitive, that is, variables *D* and *d* will be indistinguishable to VBA. VBA will make all variables of the same name the same case. For example, if you define variables *D* and *d*, VBA will treat them the same, based on which one was mentioned first in the program.

Many newer languages like C, C++, and Java are case-sensitive, that is, they distinguish between uppercase and lowercase. In such languages, *D* and *d* are different variables. Be careful of this if you are programming in languages that are case-sensitive.

TIP

Uppercase O ("Oh") is hard to distinguish from a 0 (zero), while a lowercase l ("el") looks like a 1 (one). If you wish to use O, use OO, and for l, ll, to make them more distinguishable.

2.4.2 Arithmetic Operations

VBA permits five mathematical operations. The basic arithmetic operations of addition, subtraction, multiplication, and division are indicated by $+$, $-$, $*$, and $/$, respectively. For example, $a + b$, $a - b$, $a * b$, and a / b. Raising a number to a power is indicated by a $^\wedge$ (caret): for example $2 ^\wedge 3$ cubes 2.

VBA statements are read and executed left to right. More than one statement can be placed on a single line if they are separated by a colon. For example,

```
c = a + b
d = c^2
```

or

```
c = a + b:  d = c^2
```

will add a and b to produce c and then square c to produce d. However, unless the multiple commands on a line are short, this tends to be confusing to read. Spaces do not affect what the computer does, but their use tends to enhance the clarity of the program to human eyes. Programs should always be written to be easily read (by somebody else or even by you sometime later). For that reason, avoid crowding by using generous white space.

2.4.3 Assignment Statements

Assignment statements store the result of a calculation in the location assigned to a named variable. Thus,

```
c = a + b
```

takes the values of a and b from memory, adds them, and stores the result in the location reserved for variable c. Anything previously stored as c will be replaced by this new result . Therefore,

```
c = c + a
```

adds a to the old value of c and replaces that value of c with this new result. Thus, assignment statements are different from statements in algebra. The assignment is always of this form, with the new result on the left of the equals sign. In algebra,

```
c = c + a
```

could lead to subtracting c from both sides of the equation, leaving the result that $0 = a$. A statement such as

```
a + b = c
```

is not permitted in assignment statements, another difference from algebra. Only one variable is allowed on the left of the equals sign, the new value for that variable.

2.4.4 Comments

Comments are nonexecutable lines within a program. They are most often used to include notes to the user, often indicating what the next section of the program is doing. The generous use of comments makes a program easier to understand. Comment lines are displayed in green on the VBA screen to aid recognition and indicate that such lines are not executable.

Comments in VBA can be indicated by either the letters REM (short for Remark in BASIC) or a single quote ('). Thus,

```
REM input data read in now
```

and

```
' Check to see if the velocity is positive
```

are comments.

TIP

Commenting is a handy way to temporarily deactivate lines of code from execution without actually removing them from the program. This might be used if certain output lines are sometimes desirable—for example, in checking the progress of a program during development. Simply deactivate or activate the lines by adding or removing REM or the ' that begin the lines to be deactivated or activated.

Either form may be used but REM or ' must be used on each line that is a comment; otherwise, the computer will try to execute the statement.

It is a very good idea to include brief documentation, *Type* statements (if used), and file *Opens* (if used) at the top of your program just below the defining *Sub* line. Type statements and Opens will be discussed later in this chapter.

2.4.5 Hierarchy of Operations

Calculations in VBA are performed in left-to-right order. Exponentiation is performed before multiplication and division. Multiplication and division are performed before addition and subtraction. Thus,

```
x = 2^3 + 3/4 + 7 * 8 + 9 - 6 = 8 + 0.75 + 56 + 3 = 67.75
```

Parentheses take precedence and should be used to group quantities to be included in a single operation. If parentheses are used, the number of left and right parentheses must match. This may require literally counting them. For example,

```
x = 6 + 8/2
```

yields 10 because the division is performed before the addition. But

```
x = (6 + 8)/2
```

yields 7 because the addition in the parentheses is performed before the result is divided by 2.

A common programming error is to fail to group all variables in a divisor within parentheses. Thus,

```
x = 2 + 3/(4 + 6)
```

yields 2.30, but

```
x = 2 + 3/4 + 6
```

yields 8.75. When in doubt, parentheses should be used; too many unnecessary parentheses, however, may lead to code that is hard to read.

2.4.6 VBA Library Functions

VBA contains a limited number of mathematical functions. (In Chapter 6, we shall see that these can be supplemented by importing Excel functions into VBA. A list of these is in Appendix F.) The library functions such as sin, cos, and square root are in the following table. Angles used in the trig functions must be *in radians*; the angle returned from the arctangent function is in radians.

Thus, to find the sin of one radian (about 57.3°), we could write

```
z = sin(1)
```

or

```
angle = 1
z = sin(angle)
```

(VBA does not accept Greek letters.)

Purpose	Name	Example
absolute value	Abs(x)	Abs(-6) = |-6| = 6
arctangent	Atn(x)	Atn(1.5574) = 1
cosine	Cos(x)	Cos(1) = 0.5403
exponential e	Exp(x)	Exp(3) = e³ = 20.086
natural log *	Log(x)	Log(3) = 1.0986
random number	Rnd( )	
round to n decimal places	Round(x,n)	Round(3.14159,4) = 3.1416
sine	Sin(x)	Sin(1) = 0.8415
square root **	Sqr(x)	Sqr(2) = 1.414
tangent	Tan(x)	Tan(1) = 1.5574
truncate to integer	Int(x)	Int(3.14159) = 3
x modulo y (remainder)	x Mod y	7 Mod 2 = 1

* The natural log (ln) of x in Excel spreadsheets is Ln(x).

** The square root of x in Excel spreadsheets is Sqrt(x).

Space must separate Mod from the x and y values. Before applying Mod VBA rounds x and y to the nearest integers. Thus, 15.7 Mod 3.14 would first become 16 Mod 3 and then yield 1. Since 3.14 * 5 = 15.7, 0 would be expected. To obtain the remainder of the unrounded quotient use x/y - Int(x/y).

Unfortunately, in two instances the function names in VBA are different from those in Excel. In VBA the square root of x is Sqr(x) but in Excel it is Sqrt(x). The natural log of x in VBA is Log(x), but in Excel it is Ln(x).

2.4.7 Pi Trick

Since angles used in the trig functions must be in radians but angles are often expressed in degrees, we must be able to convert between the two angle forms. Recall that 180° equals Pi radians. Then the radians value of $d°$ is $\dfrac{d * Pi}{180}$. An approximate value of Pi is well-known, but using a more correct value in a computer program could require entering a string of unfamiliar digits. Instead, we can take advantage of knowing the tangent of $\dfrac{Pi}{4}$ radians (45°) is 1. Thus, the arctangent of 1 is $\dfrac{Pi}{4}$. Then,

```
Pi = 4*atn(1)
```

This is simple and as accurate as the computer can write it.

2.4.8 Avoiding Division by Zero

If VBA attempts to perform a division by zero, execution will halt. Sometimes a division might be by zero for a pair of numbers encountered, but we do not want to halt execution because the situation is temporary. For example, in evaluating the step-size of the 4th-order Runge–Kutta method for solving ordinary differential equations in Chapter 8, we shall introduce the Collatz criterion:

$$\text{Collatz} = \left| \frac{k_2 - k_3}{k_1 - k_2} \right|.$$

At the beginning of the process, k_1 and k_2 might be equal, which would halt execution. To avoid this, we could define a very small quantity, say *tiny*, as 10^{-10}. This could be added to the denominator as

$$\text{Collatz} = \left| \frac{k_2 - k_3}{k_1 - k_2 + tiny} \right|.$$

This will not affect normal calculations, but if k_1 and k_2 are equal, *tiny* will prevent the denominator from being zero and calculation could continue.

■ 2.5 Input/Output

Obviously, getting information into and out of the computer is extremely important. We need a way to communicate with the computer. VBA provides input and output in three distinct ways: (1) in from the keyboard, out to the screen; (2) in and out from Excel spreadsheets; and (3) in and out from files. Combinations of these may be used if convenient; which one to use is the user's choice. For small amounts of data, the keyboard and screen are convenient. For large amounts of data, files are probably most convenient, especially if the input or output is shared with another user. Exchanging data with a spreadsheet is convenient if the source data are already in the spreadsheet or you wish to use Excel to further process results—for example, to make a chart.

2.5.1 Keyboard Input: `InputBox`

Suppose we want to enter a, b, and c, the coefficients of a quadratic equation,

$$ax^2 + bx + c = 0.$$

Within the program we could include the following lines:

```
a = Val(InputBox("Enter a now. "))
b = Val(InputBox("Enter b now. "))
c = Val(InputBox("Enter c now. "))
```

The text in quotation marks reminds us which variable we are entering.

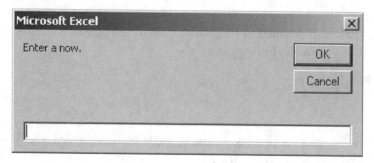

I FIGURE 2.3 InputBox.

When the program is executed, the first of these lines will produce a gray dialog box (Figure 2.3) on the screen that includes the text that was in quotes, Enter a now.

When the dialog box displays, we simply key in the numerical value of a and press "Enter". The use of Val(....) ensures that the program understands that a is a numerical value rather than the text letter *a*. We could get away with not doing this in some instances but to ensure proper use of numbers, we should include all numerical data entry via InputBox in the Val(.....) construct.

The dialog box will remain on the screen until a is entered. After this takes place the dialog box requesting the entry of b displays and, after its entry, the request for c. We can only enter one variable per line of InputBox.

The general form of the InputBox function is

```
InputBox(prompt [,title][ default][, Xpos] [,Ypos])
```

Prompt is required. Title is optional. If title is used, it puts a user-defined label at the top of the dialog box. Without specifying title, the default InputBox title is *Microsoft Excel*. The other items in brackets are optional. If we include any additional items, the place of any skipped item must be held with commas. Thus, if title is skipped but default is included, we need

```
InputBox (prompt,,default)
```

If default is used, it inserts the user-defined default value in the dialog box. If we click "OK" without changing this value, the default value is input. Otherwise, it will be overwritten and the new value will be input.

Xpos and Ypos position the InputBox on the screen. Additional parameters related to "Help" may be included to the right of Ypos.

If the following line is used in a VBA program,

```
dt = Val(InputBox("Enter stepsize", "InputBox example", 0.01))
```

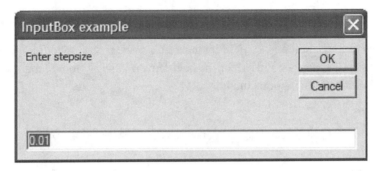

I FIGURE 2.4 InputBox with Default Value.

the dialog box in Figure 2.4 will display on the screen when this line in the program is reached.

The InputBox is labeled, the variable stepsize is prompted, and a default value of 0.01 displays. The default value will be used unless a different value is typed over it.

2.5.2 Screen Output: MsgBox

After we enter data we should output them. Outputting the values that have been input is called *echoing*. It is strongly recommended for two reasons: (1) it gives you a chance to verify that the calculation was made with the correct numbers, and (2) if printed, or stored in a spreadsheet, it provides a permanent record of the input for which these results have been obtained. Otherwise, we cannot be sure that either the input or output is correct.

Output (and messages we wish to display during execution) may be displayed on the screen using MsgBox. In the present example, we could display labeled values of a, b, and c as follows:

```
MsgBox " a = " & a & " b = " & b & = " c = " & c
```

Entries in MsgBox program lines are separated with an ampersand (&). This would produce the display in Figure 2.5.

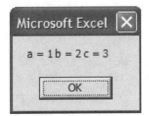

I FIGURE 2.5 MsgBox.

> **TIP**
>
> Achieve spatial separation in `MsgBox` by including space before and after the quotation marks. This was not needed for the beginning *a* label. Without labels, numbers would run together. For example,
>
> ```
> MsgBox a & b & c
> ```
>
> would run the values of a, b, and c together without spaces as 123, making interpretation difficult.

We could output each coefficient label and value on a separate output line if we had used the command `Chr(13)` to tell the computer to drop down a line for each. For example,

```
MsgBox " a = " & a & Chr(13) & " b = " & b & Chr(13) & " c = " & c
```

Alternatively, we could have put each label plus value in separate lines of `MsgBox`:

```
MsgBox " a = " & a
MsgBox " b = " & b
MsgBox " c = " & c
```

Be careful to type precisely. If we had typed `MsgBox "a ="` & b, we would have an a label but a b value. VBA cannot read!

`MsgBox` may also be used to display messages during the execution of a program. For example, we might wish to monitor the progress of a program during execution and include statements like:

```
MsgBox "About to solve for real roots."
```

or

```
MsgBox "Only complex roots for this equation."
```

The general form of the `MsgBox` function is

```
MsgBox(prompt[,buttons][,title][,helpfile,context])
```

Prompt is required. The other terms in brackets are optional. If any term is skipped to the left of those used, its place must be held by a comma. Buttons provides a choice of the buttons in the dialog box generated by MsgBox. For example, a box might contain an OK button (button = 0), an OK and Cancel button (button = 1), or a Yes, No, and Cancel button (button = 3), among others. A string of text in quotes for the title argument would put a heading on the dialog box. This could be the name of the program, for example. Without it, the default title is *Microsoft Excel*.

The value of button may be returned from using MsgBox if the MsgBox function appears to the right of an equals sign as in the following example. This returned value may then be used in making program decisions.

```
a = Val(InputBox("Enter a."))
test = MsgBox(a & "Hₐ=" Cancel program if a = 0," Otherwise OK",
vbOKCancel,      "Msgbox example")
If test = 2 Then Stop
MsgBox "a<> 0 so test = 1 and program continues."
```

In the first line, a is read. Following this, a dialog box displays (see Figure 2.6) showing two buttons: OK (button 1) and Cancel (button 2). If the OK button is selected, execution continues.

If a is not 0, execution continues, yielding the screen message shown in Figure 2.7.

A complete list of the button function labels and their numerical values may be displayed by clicking on "Help", specifying "MsgBox", and choosing "MsgBox function (Visual Basic for Applications)" from the topics list.

By the way, data using scientific notation may be entered using e for exponential: for example, $0.5e^{-06}$ instead of 0.0000005.

Formatting of numbers in MsgBox may be controlled. For example, to display x with 2 decimal places and y with 4, we would type

```
MsgBox " x = " & format(x, " .## ") & " y = " & format(y, " .#### ")
```

where the number of decimal places equals the number of number (#) signs.

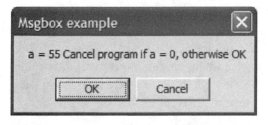

I FIGURE 2.6 MsgBox with OK and Cancel Buttons.

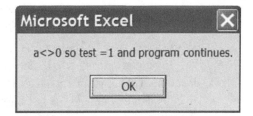

I FIGURE 2.7 MsgBox with OK Button.

2.5.3 Data from and to the Spreadsheet

Numerical data may be input from the spreadsheet using cells *(row,column).value*, where *row* and *column* denote the location of the information. *Value* is needed so that VBA recognizes the input as numerical data instead of text. This is equivalent to using `Val()` in `InputBox`. For example, suppose the value of *a* is in cell (A1), *b* in (A2), and *c* in (A3). We could access these as:

```
a = cells(1, "A").value
b = cells(2, "A").value
c = cells(3, "A").value
```

Notice that the column letter is in quotation marks.

    A is column 1 in the spreadsheet, *B* would be column 2, and so forth. Since VBA is not case-sensitive, uppercase and lowercase letters like *A* and *a* are treated the same. But it is customary to refer to Excel columns with uppercase letters. Notice that the order in VBA is *row, column*, which is the reverse of Excel where it is *column, row*. We also could have referenced the column by a numerical value. For example,

```
a = cells(1,1).value
b = cells(2,1).value
c = cells(3,1).value
```

    Output is handled by essentially reversing the two sides of the input statements. We could label the values of *a*, *b*, and *c*, with computer-generated labels, as:

```
cells( 4, 1) = "a"
cells( 5, 1) = "b"
cells( 6, 1) = "c"
cells( 4, "b").value = a
cells( 5, "b").value = b
cells( 6, "b").value = c
```

This would place the labels for *a*, *b*, and *c* in rows 4, 5, and 6 of the first column (column A) and their corresponding values in the corresponding rows of the second column (column B). Again, value is appended to the location so the cell value is interpreted as numerical rather than text. Output to cells in different worksheets, say the value of *a* to Worksheet2, may be specified as: `sheets("sheet2").cells(1,1).value = a`.

    In outputting values to a spreadsheet, we must be careful to specify the proper cell locations. Otherwise, we could accidentally overwrite values that were previously placed in cells. For example, if the last line were incorrectly typed as

```
cells( 5, "b").value = c
```

the value of *c* would replace the value of *b* just placed in cell (B5).

Labels may be included in the cells along with output by placing them in quotation marks. A label and a value may appear in the same cell by connecting them with an ampersand as

```
Cells(1, 1) = "a = " & a & " ft/s"
```

If *a* = 10.5, this would place a = 10.5 ft/s in cell (1,A).

Since the cell contains some text, the contents could not be used in spreadsheet arithmetic.

A range of a spreadsheet may be cleared by

```
Range("upperleftcell:lowerrightcell").ClearContents
```

where uppercell is the upper left corner cell of the range and lowerright is the lower right corner. For example,

```
Range("A1:C100").clearcontents
```

would clear all cells from row 1, column A, to row 100, column C.

The format in a cell may be specified using the Format menu in Excel. But it may also be controlled from within VBA. For example, to write pi to 4 decimal places, we could use the following:

```
Cells(11, 1) = Format(pi, "#.####")
```

2.5.4 Data from and to Files

Often data are stored in computer files. Before a file may be read by VBA, it must be opened and its path location specified. This is done with an Open statement. For example, if the values of *a*, *b*, and *c* are stored in a file on the F: drive called quadequationinp.txt, we could use this:

```
Open "F:quadequationinp.txt" for Input as #4
```

Both the path and the filename are within the quotation marks. If no path information is specified, VBA will look in the current directory. The current directory can be displayed in the Excel spreadsheet by typing =INFO("directory") in a vacant cell. Full path information will appear in that cell.

The number 4 following the number sign (#) in the Open statement is arbitrary, as we might arbitrarily assign a jersey number to an athlete. The number chosen must be in the range 1 to 511. It simply assigns a numerical label to the file, for use when the file is accessed within the program. If we had additional input files with different names, we would open them the same way but assign them different numbers: for example,

```
Open "F:quadequationnewinp.txt" for Input as #3
```

We access the coefficient data in the file using this statement:

```
Input # 4, a, b, c
```

The 4 tells the computer which file to access. The computer then "reads" the input in the order in which it appears in the file. The first value encountered is assigned to a, the second to b, and so on.

The input file must contain only the data to be read, and the data must be in the order in which they are read in the program. No comments can be included. Since our input invocation line had a, b, and c on one line, we should put them in a single line in the input file. Data may be separated by spaces or commas. Thus, our input file might consist of the following line:

```
1 3 2
```

or

```
1,3,2
```

We could put the data on separate lines. After reading a, the computer would drop down one line in the file looking for b, and then drop again looking for c. Thus,

```
Input #4, a
Input #4, b
Input #4, c
```

where the input file would have

```
1
3
2
```

This is less efficient than entering the input data on one line. And if a, b, and c were all on the first line in the file, b and c would not be input because the computer would have dropped to the next line in the file after reading a.

If we do not include all of the data called by `Input` in the input file, we will get an error message telling us that there is "input past end of file."

After all input data from a file have been read, the file should be closed. In our example, assuming only file #4 was used, we would type this:

```
Close #4
```

If we wish to output to a file we must also open that file, declare its path, designate it as an output file, and assign it a number.

TIP

Write `Open` statements at the beginning of the program. Open all that are needed.

TIP

It is a good idea to include `inp` or a similar designation at the end of the input file label and `out` (e.g., `quadequationout.txt`) for output files. It also makes sense to label these in the same way as the program that uses them: for example, `quadequation` (if there is only one such file). This helps us to recognize the program and its corresponding input and output files as a group. Input and output files should be created as .txt files.

Output files must be opened in the same way as input files. Path information may be different. If no path is specified, the output file will be in the current directory. To create an output file named `quadequationout.txt` in directory F and designate it as file 7, we could type this:

```
Open "F:quadequationout.txt" for output as #7
```

We could then echo our input to this file and later output our results using statements like the following:

```
Print #7, "a = ", a, "b = " , b, "c = " , c
    .....
Print #7, " x1 = " , x1, " x2 = " , x2
```

`Print #7` will send this output to the file designated as 7. Instead of `Print`, we could have used the keyword `Write`. However, `Print` gives more reasonable spacing along the output line than does `Write`.

When completely finished with this file, we must close it accordingly:

```
Close #7
```

Failure to close the file would prevent us from viewing it. If not closed, the file is open and still active, waiting for further action until we close either it or the Excel file. Closing Excel would be inconvenient because we often want to look at file results before closing the program.

If we use the same number for both input and output files, when output is invoked it would go to the next line in the file after the last one read or written to. This usage is less common and potentially troublesome. Usually we use separate input and output files.

Printed output may also be formatted. For example, to output Pi (already defined) to 4 decimal places we use

```
Print #7, format(Pi, "#.####")
```

which would yield 3.1416. The number of #'s indicates the number of places left and right of the decimal place.

To print in scientific notation we could use

```
Print #7, format(Pi, "#####E##")
```

which would yield 31415E-4. Here the number of #'s indicates the number of places left and right of E.

Alternatively we could use

```
Print #7, format(Pi, "scientific")
```

which yields 3.14, the result always with 2 decimal places.

2.5.5 Hardwired Data

It is not a good idea to hardwire the values of variables into the program instead of inputting them: for example,

```
a = 1: b = 2: c = 3
```

However, it is tedious to input data repeatedly during the development of a program. In this situation, it makes sense to temporarily include fixed values of the test input within the program, without having to enter these each time the program is run during development. Once the program is running correctly, a more convenient means of input can be activated.

■ 2.6 A Simple Program

Let us now write a simple program to add two numbers together and display the result on the screen. At first, we will define the numbers to be added within the program; later, we will enter them from the keyboard.

```
[PROGRAM PROG1]

Sub Prog1()
REM add values of a and b to produce result c
REM a and b defined within the program
REM answer is displayed on the screen along with the echoed
values of a and b
    a = 2: b = 3
    c = a + b
MsgBox "a = " & a & " b = " & b & " c = " & c
End Sub
```

After the Sub heading, we define the values of *a* and *b*. The next line computes the sum and stores it in the location for *c*. Next, the values of the inputs *a* and *b* are echoed to the screen and their sum, *c*, is output. Notice that each of these numbers is labeled.

If we wanted to (1) input the value of *a* via InputBox, (2) input the value of *b* from the spreadsheet, and (3) echo both these and the results to the spreadsheet, we would write our program as follows:

```
[PROGRAM PROG1]

Sub Prog1()
' add values of a and b to produce result c
' a input via InputBox, b from the spreadsheet
' the answer is sent to the spreadsheet along with the echoed
' values of a and b
    a = Val(InputBox("Enter a."))
    b = cells(1,2).value
    c = a + b
    cells(2,1) = "a"
    cells(2,2).value = a
    cells(3,1) = "b"
    cells(3,2).value = b
    cells(4,1) = "c"
    cells(4,2).value = c
End Sub
```

Notice that we have included labels for the results to be output to the spreadsheet. We also had to be careful that we did not try to put two different quantities in the same spreadsheet cell. If we had tried to enter two quantities in the same cell, only the last one written would display. Previous entries would be overwritten.

■ 2.7 Documentation

Program documentation is an essential part of producing a good working program. For a simple program, it is usually sufficient to include the documentation in the program itself using comment lines at the top of the program. This has the advantage that the documentation is integral to the program, which prevents it from being misplaced.

Programs associated with this book can be documented using comments. More complicated programs may require a separate user's manual to explain all the features of a program that are needed for successful use.

At a minimum, documentation should describe what the program does, list the input variables, and list the output results. If input variables have units, these should be specified. Otherwise, users would not know if the temperature input should be in degrees Fahrenheit or Celsius, for example. Any restrictions on the program should also be mentioned. Examples might be that the program only finds real roots or that it is limited to solving 75 linear equations. The date of the last revision and the programmer's name(s) should also be included.

The program in Section 2.6 illustrates the use of simple documentation. Another example, at the top of the program in Appendix H, is for a program written to solve a pair of ordinary differential equations using the Cash–Karp method. That documentation is reproduced here for convenience.

```
                        *** FILENAME CKODE ***

'   LAST MODIFIED 11/22/07
'   TO SOLVE TWO ORDINARY DIFFERENTIAL EQUATIONS USING
'   THE CASH-KARP (RUNGE-KUTTA-FEHLBERG) METHOD. INPUT ARE
'   THE INITIAL AND FINAL INDEPENDENT VARIABLES, INITIAL
'   DEPENDENT VARIABLES, STEP-SIZE, PRINT-FREQUENCY, AND
'   ERROR TOLERANCE.
'   THESE ARE IN THE SPREADSHEET. OUTPUT IS TO THE
'   SPREADSHEET AND A FILE.
'   X IS THE INDEPENDENT VARIABLE; Y AND Z ARE THE
'   DEPENDENT VARIABLES.
'   THE DERIVATIVES ARE DEFINED IN FUNCTIONS F (DY/DX)
'   AND G (DZ/DX).
'   THE ERROR IS CHECKED DURING PRINT-OUT. IF THE ERROR IS AN
'   ORDER OF MAGNITUDE DIFFERENT FROM THE TOLERANCE THE
'   STEP-SIZE IS ADJUSTED.
```

■ 2.8 Running VBA

Once a VBA program has been written and the syntax errors corrected, it is ready to be run. We do this by (1) clicking on the delta (▶) on the toolbar above the program listing, or (2) choosing "Run" from the Toolbar menu and selecting "Run" or pressing F5. The VBA module screen for the program in Section 2.6 displays as shown in Figure 2.8. A VBA program also can be run from the Macro Box called from Tools/Macro/Macro above the Excel spreadsheet by clicking the "Run" button.

If the program encounters any errors, it will stop at the first error. The error must be corrected before the program will proceed. You can continue the program then from this point by pressing the delta again or selecting "Continue" from the Run menu. In this way, the program will run from one error to the next until all have been located and eliminated. This does not mean the results are correct; it simply means that the program has consistent syntax and variables to get through a run. Results should always be compared to known results the first time a program is run.

▌ FIGURE 2.8 VBA Module Screen.

TIP

Sometimes it is obvious that a running program simply is not working. Perhaps an iterative solution is not converging but is running endlessly. To halt execution, press the "Control" and "Break" keys together. These are the lower-left and upper-right keys on the keyboard.

■ 2.9 Flowcharts

Flowcharts are an outline for a program, similar to the outline for an English paper. Emphasis is on the program's logic and how the various parts fit together. A flowchart should be generic: that is, independent of the programming language to be used. Thus, the same flowchart could be applied to a program in VBA, Fortran, C, or C++, for example. If the flowchart is written correctly, the program itself is much easier to write, for it requires only using the language syntax to implement the logic already developed. The use of a flowchart, therefore, separates the two areas of difficulty in programming: the program logic and the language syntax.

The basic symbols used in flowcharting follow. Flowchart symbols are connected with arrows that show the direction in which the program proceeds. The basic flowchart symbols are shown in Figure 2.9.

A flowchart for Program 1 (Section 2.6) is shown in Figure 2.10.

An oval is used for starting and ending programs.

A rhombus is used for input and output.

A rectangle is used for calculations.

A diamond is used for decisions.

A hexagon is used to delineate loops.

I FIGURE 2.9 Basic Flowchart Symbols.

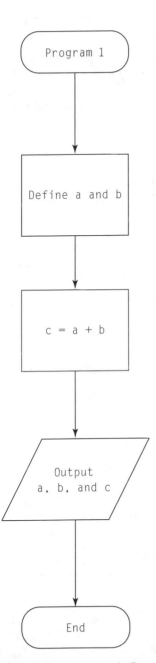

I FIGURE 2.10 Flowchart for Program 1.

■ 2.10 Variable Types

VBA is not a strongly typed language. Unlike some other languages, variables need not be typed before being used. Variables that are not typed are called *variants*. They take on the characteristics of a *double*, which is 15 significant figures for decimal numbers. Besides variants, the most common variable types used in science and engineering calculations are as follows:

Integer	for integers
Single	for decimal numbers, 6 significant figures
Double	for decimal numbers, 15 significant figures
String	for alphanumeric (text) characters

Because other high-level languages like Fortran and C/C++ use variable typing, the practice in this book will be to type all variables. Again, this is not necessary in VBA, but it will help the programmer to be more aware of the use of defined variables and to help make a transition to other languages.

We can require programs we write to use typed variables if we select "Require Variable Declaration" under Tools/Options from the pull-down VBA menu. This places *Option Explicit* at the top of a new program when it is written. Execution will halt when an untyped variable is encountered.

We type variables with

```
Dim variable name as type e.g.

Dim x as single
Dim y as double, z as double
Dim i as integer
Dim month as string
```

Each variable must be typed separately, unlike other higher-level programming languages. Thus, *y* and *z* were separately typed following `Dim`, even though both are doubles.

TIP

Type all variables used in the main subprogram at the beginning of that subprogram, perhaps right after any `Open` statements. They need be typed only once in a program, unlike most other languages.

Variables introduced in later subprograms or functions should be typed there. (See Chapter 6.)

VBA takes typed variables seriously. If *d* is typed as an integer, "*d* = 1/3" would yield 0 as the result since its value is less than 1. Similarly, "*d* = 7/3" would yield 2 as the result.

VBA data types, their storage requirements, and their range of values follow:

Integer	2 bytes	−32,768 to 32,767
Long	4 bytes	−2,147,483,648 to 2,147,483,647
Single	4 bytes	−3.402823E38 to −1.401298E-45 for negative nos.
		1.401298E-45 to 3.402823E38 for positive values
Double	8 bytes	−1.79769313486231E308 to
		−4.94065645841247E-324 for negative values,
		4.94065645841247E-324 to
		1.79769313486232E308 for positive values

2.10.1 String Variables

String variables are text variables. They may be used for labels or in records as names. If we had typed *month* as a string, we could give it a value by reading it in or typing, for example:

```
month = "January"
```

Two convenient string variables available from the VBA library are `Time$` and `Date$`. The $ on the end indicates they are string variables. If `Time$` is output, it provides a record of the computer system clock at the time it was executed. Likewise, `Date$` provides the current date. Both of these are useful in printed output, for they document when the particular program run was made.

2.10.2 Inputting a Filename

The name of an input or output file may be input with `InputBox` during execution. This means the program does not have to be changed each time a different data file is read by the program. This is accomplished by using a string variable. In the following lines of code, the variable filename, `ffnamein`, is input via `InputBox` and the named file is then opened. Of course, the named input file must exist. Path information may be included in the filename. For example, if the input file is on a flash drive identified by the computer as the letter F, the filename typed in response to the `InputBox` prompt might be `F:quadequationinp.txt`. Output filenames can be read in the same way.

```
Dim ffnamein As String
ffnamein = InputBox("Enter ffnamein. ")
Open ffnamein for Input As #4
```

■ 2.11 Example: The Real Roots of a Quadratic Equation

Suppose we want to compute the real roots, r_1 and r_2, of a quadratic equation of the form

$$ax^2 + bx + c = 0.$$

These roots are given by the well-known equation

$$r_{1,2} = \frac{-b \pm \sqrt{(b^2 - 4 * a * c)}}{(2 * a)}.$$

A flowchart for a simple program to perform this calculation is shown in Figure 2.11, followed by the program itself. The program, as written, will only work for real roots. If the discriminant, the $b^2 - 4 * a * c$ that is under the square root, is negative, execution will be stopped by the computer, which cannot take the square root of a negative number. This would constitute a computer "bug," that is, an execution error resulting from the programmer incorrectly anticipating a possible result of the calculation. In Chapter 4, we will correct this possible bug when we examine *decisions* and *branches* made within a program.

[PROGRAM QUAD]

```
Sub Quad()
' A program to compute the real roots of a quadratic
' equation, ax² + bx + c.
' These coefficients are prompted as input from the
' keyboard. The discriminant is calculated but no decision
' made based on its value. Hence execution will halt and
' an error will result if the square root of a negative
' discriminant is attempted. Otherwise the two real roots
' r1 and r2 will be computed and output to the screen.

    Dim a, b, c, d, r1, r2 as double

a = Val(InputBox("Enter a"))
b = Val(InputBox("Enter b"))
c = Val(InputBox("Enter c"))
MsgBox "a = " & a & " b = " & b & " c = " & c
d = b*b - 4*a*c
MsgBox " discriminant = " & d
r1 = (-b + sqr(d))/(2*a)
r2 = (-b - sqr(d))/(2*a)
MsgBox " root 1 = " & r1 & " root2 = " & r2
End Sub
```

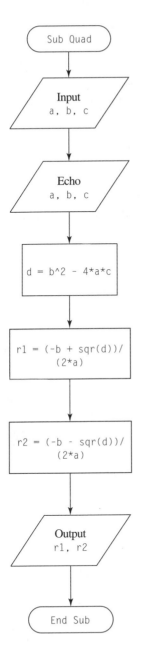

Notice the following:

1. Brief documentation is provided at the beginning of the program. Variables are typed.

2. Then *a*, *b*, and *c* are prompted and input via InputBox. This method is arbitrary but convenient for small amounts of data.

3. Labeled input values are displayed (*echoed*) to the screen for verification.

4. The discriminant, the value of $b^2 - 4 * a * c$, is then specifically computed for two reasons. The second reason is to avoid having to compute it twice in the lines that compute r1 and r2. But the more important reason is something not actually used in this program. The sign of the discriminant should be checked for the possibility of a complex solution. This involves making a decision, which will be covered in Chapter 4. For the time being, *d* is merely calculated. If *d* is negative for the present program, an error message will be generated and execution will stop.

5. The VBA library function, sqr, for the square root is used. (A list of VBA library functions was presented in Section 2.4.6.) Compared to other languages, the number of VBA library functions is small. Later, we will see that we can include Excel spreadsheet functions in VBA programs.

6. The roots, r1 and r2, are computed.

Check data should be run with this program. If a = 1, b = 3, and c = 2, then r1 = −1 and r2 = −2.

■ 2.12 User-Defined Types—Complex Variable Type

VBA allows the user to create new variable types. User types begin with the word *Type* and end with the words *End Type*. They must be defined ahead of all Subs and functions; thus, they are at the top of the program. Typically, they may be used to define a variable that has several members or attributes. For example, a complex variable would have a real and an imaginary part.

If we wanted to define a complex variable, we could have

```
Type complex
     real As Double
     imag As Double
End Type
```

at the top of our program. If, subsequently, $A = 2 + 3i$ is a complex variable, we could write it in VBA as

```
A.real = 2
A.imag = 3
```

where .real indicates the real part of *A* and .imag is the imaginary part.

The following program segment shows how two complex variables, *A* and *B*, could be added in a program to yield *C*. *A* and *B* must be typed before they are used.

```
Type complex
     real As Double
     imag As Double
End Type
Sub complexadd()
Dim A as complex, B as complex, C as complex
A.real = 2
A.imag = 3
B.real = 4
B.imag = -5
C.real = A.real + B.real
C.imag = A.imag + B.imag
MsgBox "C = " & C.real & " " & C.imag & " i "
  . . .

  . . .
End Sub
```

Other uses of user-defined types might include characteristics of a product, such as the following:

```
Type product
     Color as string
     Weight as double
     Serial number as long
End Type
```

More than one user-defined type may be used in a program. A user-defined type could also use another user-defined type.

■ 2.13 Debugging

Debugging your program requires a combination of skill, insight, patience, and some-times a little luck. The process is greatly simplified by intelligently interpreting com-piler error messages, the use of extra, diagnostic output statements, and the debug features of VBA. These function like X-rays to locate internal problems. For example, if you believe that variable A causes division by zero, display or write out all variables that are used to determine the value of A. Computer-generated syntax error messages will cause the error line to be displayed in red. If the error occurs during execution, a gray dialog box will display. Selecting "Debug" will identify the error line.

If no error message results but answers are incorrect, compare key variables in a cal-culation with hand calculations of the same variables for at least the first iteration. Usu-ally, one iteration is not so complex that this becomes too tedious.

Three common mistakes in writing programs that may be hard to spot are as follows:

1. Interchange of zero (0) and the letter O ("oh"), or the number 1 and lowercase letter L.

2. Inconsistent use of parentheses.

3. Arrays too small, inconsistent, or not declared. (See Chapter 10.)

2.13.1 A Debugging Example

Suppose in running the program Quad, discussed in Section 2.11, we had accidentally entered a value of 0 for a, using InputBox. When the program got to the line where r1 was calculated, a division by 0 would occur. Because VBA will not permit this, an error message would appear in a gray dialog box when the program reached this line. It would say "Runtime error." If we click on "Debug" at the bottom of the box, the computer will highlight the line where the error occurred in bright yellow. If we drag the cursor slowly over the line, the various values will appear in pale yellow. We will soon see that a is 0, the cause of our problem.

2.13.2 VBA Debugging Tools

VBA has several helpful program tools under the Debug pull-down menu, shown in Figure 2.12.

I FIGURE 2.12 Debug Tools.

When "Run To Cursor" is selected, the program will run to the location of the cursor in your program, but without executing that line. If you slowly drag the cursor over any variable, the current value of that variable will be displayed. Variables not yet given a value will be labeled *Empty*.

"Toggle Breakpoint" works similarly. You can set multiple *breakpoints* (stopping points) in your program. Click in the gray left margin opposite the line. Breakpoints are indicated by a brown circle in the left margin of the program. Active breakpoints can be removed by clicking again in the left margin. During execution, the program stops at each breakpoint. There you can view the value of variables existing prior to the execution of that line. If you want to see values after a line execution, select the next line as the breakpoint. As an alternative, insert a dummy line right after this one (like `cat = 6`) and make that a breakpoint. To continue execution after the program has stopped at a breakpoint, (1) use the delta (▶) on the toolbar, or (2) select "Continue" from the Run menu, or (3) press F5. (NB: Continue is an option if the program has stopped at a breakpoint or selected step.)

"Add Watch" lets you select variables in advance whose current values at a point in the program will be displayed. You must enter these variable names. "Quick Watch" will also keep track of variables. It identifies them as the variable left of the equals sign (i.e., the assignment value) for the line where Quick Watch is selected.

Selecting "Step Into" (or pressing F8) lets you step through a subprogram one line at a time. Thus, variables can be viewed as the program progresses. "Step Over" executes code one procedure, or statement, at a time. "Step Out" executes the remaining lines of a procedure in which the current execution point lies.

In this way, the computational history of variables can be viewed. If you have a more complicated program, you might want to add additional `Print` statements to your program to keep track of certain variables and print them to an output file.

■ 2.14 File Saving and Security Level

VBA programs are automatically saved when you save the spreadsheet in which they were generated. Subsequently, opening a VBA program depends on the security level under which it was saved. Access to macros is controlled through a dialog box (see Figure 2.13) that lets you decide if the macro should be opened when the file is opened. This provides both security and convenience. Clicking on "Enable Macros" allows the file to be opened.

The security level is controlled under the Excel Tools/Macro pull-down menu. From it, select "Security". This yields the dialog box shown in Figure 2.14. Specify the desired security level for the spreadsheet. "Medium" is a good choice. It allows you to choose whether you wish to run the macro, thus improving security, without the complications of the higher security levels.

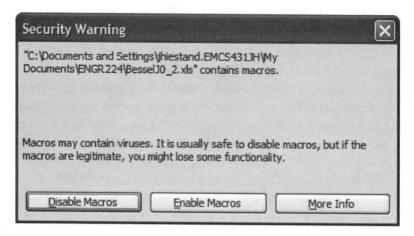

I FIGURE 2.13 Security Warning Dialog Box.

The security level must be set before an attempt is made to open a file. If access is denied because the security level is too high, reducing the previous security level does not take effect until the next time Excel is opened.

■ 2.15 A Word of Encouragement

Programming is both fun and tedious. The fun occurs when you successfully run a program to compute something that would be prohibitively time-consuming to do by hand. In this sense, writing a good program is like solving a puzzle.

The tedium occurs when trying to debug a program. Programmers should remember that computers are obedient but stupid! They do exactly what we tell them, but we

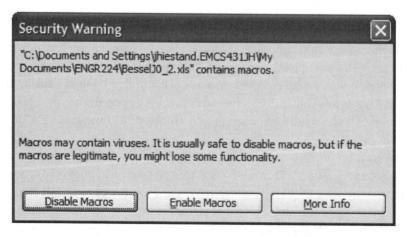

I FIGURE 2.14 Security Choices.

must be very precise. We must use their vocabulary and "speak" in precise sentences. It is like talking to a well-behaved pet that is much more obedient than a child with a larger vocabulary. The error messages provided for syntax errors are invaluable for debugging language errors.

For debugging logical errors, outputting key variables as the program proceeds is essential (it is a bit like medical imaging is used to see what is inside the human body). Outputting key variables provides a way to check the program's progress and verify that the program is "on course" or identify where it went awry.

And when you have put forth a good effort and are still stuck, get away from your computer for a while. Our minds quickly get into ruts and we begin to overlook mistakes. Go do something else. Often, the mistakes will be obvious, or nearly so, when we return. When difficulties are encountered, two shorter sessions are probably more efficient than one long session of total equivalent length.

Persevere; it is worth it. And enjoy.

■ 2.16 Chapter 2 Exercises

2.1 Write a program to input the coefficients of a quadratic equation and solve for the roots. Apply your program to the equation

$$x^2 + 2x - 3 = 0.$$

Output your labeled input and labeled results.

2.2 Temperatures may be converted from Celsius to Fahrenheit using the formula F = 1.8C + 32. Write a program to input the temperature in Celsius and convert it to Fahrenheit. Try your program with C = −40, C = 0, C = 20, and C = 100.

Input Celsius. Output your labeled input and labeled results.

2.3 A meter equals 3.281 feet, a mile is 5280 feet, a minute contains 60 seconds, and an hour contains 60 minutes. Write a program to convert miles per hour to meters per second.

a. Input the four individual conversion factors.

b. Echo labels and their values to a spreadsheet.

c. Output labeled results to both screen and spreadsheet.

2.4 The cosine of an angle in radians can be approximated by a Maclaurin series as

$$1 - \frac{x^2}{2!} + \frac{x^4}{4!} - \frac{x^6}{6!} \cdots$$

where x is in radians.

Begin with a flowchart. Write a program to approximate the cosine through these first five terms. Use it to estimate the value of the cosine of 34°. Enter the angle in degrees and convert it to radians within your program. Also compute the exact answer using the VBA library function for cosine, cos(x). Output the series answer, the last series term computed, the exact answer, and the difference between the approximate (series) and exact answers.

2.5 Given two points with coordinates (x_1, y_1) and (x_2, y_2), the length of the line connecting them is given by

$$ll = \sqrt{(x_2 - x_1)_2 + (y_2 - y_1)^2}$$

and the slope of that line by

$$m = \frac{y_2 - y_1}{x_2 - x_1}.$$

If point 1 is at $(1, 3)$ and point 2 is at $(6.5, -8)$, compute the length of the line between them and its slope.

2.6 Given the length of the hypotenuse and one of the angles less than 90° in a right triangle, compute the other two sides. Verify these are correct by using Pythagoras' theorem to compute the hypotenuse. Let $h = 6$ and $\Theta = 37°$.

2.7 Polar coordinates are related to Cartesian coordinates by the equations

$$x = r * \cos(\theta) \quad y = r * \sin(\theta) \, r = \sqrt{x^2 + y^2} \quad \theta = \tan^{-1}\left(\frac{y}{x}\right).$$

Write a program to input $x = 2.5$ and $y = 7.1$ to compute r and θ. Then input $r = 5.5$ and $\theta = 49°$ to compute x and y.

2.8 The law of sines is

$$\frac{a}{\sin(A)} = \frac{b}{\sin(B)}.$$

Write a program to compute the length of side b if $a = 25$, $A = 35°$, and $B = 110°$.

2.9 The law of cosines is $c^2 = a^2 + b^2 - 2 * a * b * \cos(C)$. Write a program to compute the length of side c if $a = 85$, $b = 38$, and $C = 120°$.

2.10 The sine of the sum of two angles is given by
$\sin(A + B) = \sin(A) * \cos(B) + \sin(B) * \cos(A)$. Write a program to compute the sin of $(A + B)$ using this formula if $A = 37°$ and $B = 15°$. Add the two angles and compute the sin directly. Compare your answers.

2.11 The cosine of the sum of two angles is given by
$\cos(A + B) = \cos(A) * \cos(B) - \sin(A) * \sin(B)$. Write a program

to compute the cosine of $(A + B)$ using this formula if $A = 37°$ and $B = 15°$. Add the two angles and compute the cosine directly. Compare your answers.

2.12 The area of any triangle can be calculated in terms of the lengths of its three sides according to the following formulas (a, b, and c are the side lengths).

$$s = 0.5 * (a + b + c)$$
$$\text{Area} = \sqrt{(s * (s - a) * (s - b) * (s - c))}.$$

For example, a right triangle with base 3, height 4, and hypotenuse 5 would yield

$$s = 0.5 * (3 + 4 + 5) = 6$$
$$\text{Area} = \sqrt{(6 * (6 - 3) * (6 - 4) * (6 - 5))} = 6.$$

Begin with a flowchart. Write a program with input a, b, and c, calculate the area, and output the input and the results. Apply your program to $a = 15$, $b = 27$, and $c = 35$. Echo your input and output your answer.

2.13 Some of the equations of projectile motion (without drag) follow. Vx and Vy are the x and y components of the velocity, Θ is the angle of the trajectory, subscript 0 denotes initial conditions, and t is the time.

$$Vx = V_0 * \cos(\Theta_0), \qquad Vy = V_0 \sin(\Theta_0) - g * t, \tan(\Theta) = Vy/Vx,$$
$$x = x_0 + V_0 * \cos(\Theta_0) * t \qquad y = y_0 + V_0 \sin(\Theta_0) * t - .5 * g * t^2.$$

Begin with a flowchart. Write a program to input $V_0 = 80$ ft/s, $\Theta_0 = 61°$, and $t = 1.2$ s. Within your program, convert Θ_0 to radians. Output the input and your results. The library function for sine is $\sin(x)$, for cosine $\cos(x)$, and for the arctangent $atn(x)$. Let x_0 and $y_0 = 0$.

2.14 The van der Waals equation of state for gases is given by

$$P = \frac{RT}{v - b} - \frac{a}{v^2}.$$

Write a program to input the values of R, T, a, and b, and then compute the value of P.

For air,

$$a = 343 \text{ atm} \frac{\text{atm} - \text{ft}^6}{(\text{lbmol})^2}, b = 0.585 \frac{\text{ft}^3}{\text{lbmol}}, \text{ and } R = 0.730 \frac{\text{atm} - \text{ft}^3}{\text{lbmol}°R}.$$

Apply your program to $T = 1000°R$ and $v = 5.36$ ft^3/lbmol.

2.15 Make a flowchart showing the route from your residence to school. Be sure to include all turns.

2.16 Let $A = 11 - 6i$ and $B = -2 + 5i$. Create a user-defined type for complex variables. Compute $A + B$, $A - B$, $A * B$, and A/B.

2.17 Let $\overline{V}_1 = 2\overline{i} + 3\overline{j} + 4\overline{k}$ and $\overline{V}_2 = -10\overline{i} + 20\overline{j} + 9\overline{k}$. Create a user-defined type for a vector. Compute the vector product $\overline{V}_1 \times \overline{V}_2$.

2.18 The steady-state current in an alternating L-C (inductor-capacitor) circuit as a function of time is given by

$$i = \frac{E}{\sqrt{\left[R^2 \mp \left(\omega L - \dfrac{1}{\omega C}\right)^2\right]}}$$

where E is the voltage, R the resistance, L the inductance, C the capacitance, and ω is defined in 2.19.

Write a program to input E, R, L, and C, and then compute i. Let $E = 8$ v, $R = 90$ ohms, $L = 0.25$ henry, and $C = 10^{-5}$ farad.

2.19 The current in an L–C circuit as a function of time is given by

$$i = \frac{E}{\varpi L} e^{\alpha t} \sin(\varpi t) \text{ where } \varpi = \sqrt{\left(\frac{1}{LC} - \frac{R^2}{4L^2}\right)} \text{ and } \alpha = -\frac{R}{2L},$$

provided the argument of the square root is positive.

Write a program to input t, E, R, L, and C, and compute i. Let $t = 0.001$ s, $E = 8$ v, $R = 90$ ohms, L $= 0.25$ henry, and $C = 0.000001$ farad.

2.20 The non-dimensional temperature distribution in a fin of uniform cross section with a tip temperature T_L is given by

$$\frac{T - T_{ref}}{T_b - T_{ref}} = \frac{\dfrac{T_L - T_{ref}}{T_b - T_{ref}} * \sinh(mx) + \sinh(m(L - x))}{\sinh(mL)}$$

where $m^2 = \dfrac{4 * h}{kD}$.

D is the fin diameter, h is the heat transfer coefficient, L is the fin length, and k is its thermal conductivity. Calculate T at $x = 0.25, 0.5,$ and 0.75 when $D = 0.167$ ft, $L = 1$ ft, $T_b = 250°R$, $T_{ref} = 75°R$, $T_L = 100°R$, $h = 100$ BTU/(h – ft^2 – °R), and $k = 15$ BTU/(h – ft – °R).

3 Errors, Series, and Uncertainty

■ 3.1 Types of Errors

Errors in numerical analysis result because we are making approximations. If we are approximating a noninteger, we have to write it to a finite number of places: for example, we may write $\frac{1}{3}$ as 0.33 or 0.3333 or 0.333 333 333 333, but at some point we are going to stop writing 3's and thus slightly underrepresent the exact value. The opposite occurs with $\frac{2}{3}$ when eventually we round up the last 6 to a 7 and thus slightly overrepresent the quantity. These differences constitute *round-off error*.

On the other hand, if we are evaluating a series, we may take 1 or 5 or even 100 terms, but eventually we terminate the approximation process. This termination produces *truncation error*.

Both types of error are inevitable in numerical analysis. We are always dealing with approximations. We do what we can to control errors, but we cannot eliminate them. Note that these errors are not mistakes. Saying $2 + 3 = 4$ is simply wrong (a mistake) no matter how many decimal places we might take. It is a blunder.

The *true error* (E_T) is the difference between the correct and the approximate values, written in that order:

$$E_T = \text{correct value} - \text{approximate value}.$$

Usually, we are most interested in the magnitude of this difference and so take the absolute value:

$$E_T = |\text{correct value} - \text{approximate value}|.$$

Because of round-off error, an exact result such as $\frac{1}{3}$ written to a finite number of decimal places is only an approximation. Recognizing this possibility, we shall refer to values as correct, rather than exact, in this text.

In general numerical applications, we do not know the correct value or we would use it immediately. Therefore, we must seek an alternative in estimating errors. In an iterative process, we frequently use the difference between successive values. This is called the *approximate error*, E_A:

E_A = current value − previous value.

Stated in absolute value:

E_A = |current value − previous value|.

We assume the most recent value is the more correct one, and hence the proper order is the current value minus the previous value. Again, we usually are interested only in the magnitude of this change and so we take the absolute value of the difference. If successive values get closer together, the process is *converging*. Lacking the correct value or another error estimate, E_A is often the best we can do. It is usually an accurate indicator of progress.

Relative errors are an alternate form for both the true error, E_T, and approximate error, E_A. Frequently, it is more meaningful to compare the error to a measure of the value we are seeking, either the correct value or the most recent value. If we divide E_T by the correct value, we obtain the *true relative error*, ε_T:

$$\varepsilon_T = \left| \frac{correct\_value - approximate\_value}{correct\_value} \right| * 100.$$

Again, we have taken the absolute value. We have further multiplied by 100 to express the answer as a percentage. It is important to note that since relative errors are ratios, they are dimensionless. E_T and E_A will have whatever units are in the quantities being calculated, but ε_T and ε_A will not.

The approximate relative error, ε_A, is defined similarly:

$$\varepsilon_A = \left| \frac{current\_value - previous\_value}{current\_value} \right| * 100.$$

Alternately, this may be written as:

$$\varepsilon_A = \left| 1 - \frac{Old}{New} \right| * 100$$

where *New* is the current value and *Old* is the previous one.

Should we consider absolute or relative errors? Since relative errors compare the error or change to a measure of the value, it is frequently more meaningful. For example, if you and Bill Gates both have a $100 error in your respective checkbooks, it certainly would be worth your time to find it. The relative error in his checkbook would be very small; it would not be worth his time to find it. Each checkbook would have the same discrepancy, but their relative values would be vastly different.

3.1.1 Significant Figures and Tolerance

The number of significant figures in an answer may be related to the relative error according to the following formula:

$$\varepsilon = 0.5 * 10^{(2-\text{sf})}$$

where sf is the number of significant figures and ε is a percentage. If a decimal error is calculated, the exponent is simply $-$sf. Thus, if an answer to three significant figures is desired, a relative error less than or equal to $0.5 * 10^{(2-3)} = 0.05\%$ is required. Such limits define a *numerical tolerance*.

In iterative calculations, it is customary to establish a tolerance and repeat the calculation until the tolerance is satisfied or a maximum number of iterations is performed. The latter alternative should be included to assure that the process stops, even if it does not converge to the tolerance.

Either E_A or ε_A could be compared to the tolerance. An iterative process to calculate a temperature might be continued until a temperature difference of less than one-half degree has been achieved or the percentage change was less than a certain amount. A nonrelative comparison will include units if they occur in the variable being calculated.

3.1.2 A Computer Round-off Example

A simple example of round-off problems in computers is provided by the following program segment. Single-precision numbers were used:

```
Dim x1 as single, x2 as single, x3 as single
x1 = 12345678
x2 = 98765432
x3 = x1 + x2
```

The correct value of this sum is $111,111,110$. But when executed in VBA, the displayed value of x3 was $111,111,112$. Why? Because single-precision numbers are accurate only to seven significant figures. Numbers beyond that are unreliable. The correct answer is easily obtained by declaring x1, x2, and x3 to be double. This is a good argument for using double-precision variables.

■ 3.2 Why Series?

Polynomial series are often used in approximating complicated functions. Sometimes series are used to obtain approximate closed-form solutions to ordinary or partial differential equations. Even though numerical methods and computers provide a powerful alternative to series solutions, series often are useful. They provide a standard for comparing alternate numerical schemes. For example, techniques for numerically solving ordinary differential equations are usually compared with reference to their agreement with Taylor series, as we shall see in Chapter 8. Series are also useful for approximating transcendental terms in nonlinear equations for the purpose of estimating starting values. This will be investigated in Chapter 7, Section 7.4.

■ 3.3 Taylor Series

A *Taylor series* approximates the value of a function at a point x in terms of the function and its derivatives of the function at another point x_0. A one-derivative Taylor series would be like estimating the temperature at noon in terms of the known temperature at 11 a.m. and adding a few degrees for an expected increase. Mathematically, we can make this process more precise by taking additional terms in the series. Intuitively, the closer together are x and x_0, the better the agreement will be. Practically speaking, a Taylor or any other series becomes less useful if we have to evaluate a lot of terms.

The general form of a Taylor series for $f(x)$ is

$$f(x) = f(x_0) + f^{\mathrm{I}}(x_0) * (x - x_0) + \frac{1}{2!} * f^{\mathrm{II}}(x_0) * (x - x_0)^2$$
$$+ \frac{1}{3!} * f^{\mathrm{III}}(x_0) * (x - x_0)^3 + \ldots \frac{1}{n!} * f^{\mathrm{n}}(x_0) * (x - x_0)^{\mathrm{n}} + Rn \tag{3.3.1}$$

where the superscripts on f (I, II, etc.) are the successive derivatives of f, and Rn is the remainder term. (Both x and x_0 *must* be in radians if a trigonometric function is being approximated.)

Figure 3.1 illustrates a straight-line projection (one-derivative term) from x_0 to x^*. Clearly, more terms (and higher derivatives) will yield a better result.

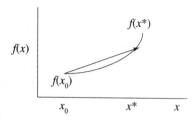

I FIGURE 3.1 Straight-line Projection of $f(x)$ from x_0 to x^*.

The form of the remainder term, Rn, is

$$Rn = \frac{1}{(n + 1)!} * f^{n+1}(z) * (x - x_0)^{n+1} \tag{3.3.2}$$

The expression $f(x)$ minus the terms through the nth derivative on the right-hand side of (3.3.1) leaves only Rn on the right side. Hence, Rn is the difference between the correct and approximate value of the series. If Rn could be computed correctly, the series would be exact. In general, we can only estimate Rn. Hence, it becomes the approximate error of the series, E_A.

The expression Rn looks like the next term in the series except that the $(n + 1)^{st}$ derivative is evaluated at some point z such that $x_0 <= z <= x$. Unfortunately, we do not know, in general, what the value of z should be in order to make the result correct. Therefore, we will do something very common, useful, and safe in numerical analysis: we will choose the location of z in order that our estimate will yield the maximum value of the derivative (in magnitude) in the range $x_0 <= z <= x$. Thus, our estimate will be conservative: this process is called *bounding the error*. If you do not know exactly how long it will take to reach a location where an event will occur with a specific starting time, you will deliberately estimate a longer rather than a shorter travel time to ensure that you will not be late.

Estimating high usually is not too hard to do. However, if the $(n + 1)^{st}$ derivative has an extreme value between x_0 and x, the location for z is harder to find. To locate it precisely, the next derivative would have to be calculated, set equal to zero, and the resulting equation solved for the location of the extreme value. This potentially tedious process can be avoided if a few trial values of the $(n + 1)^{st}$ derivative are calculated to estimate the extreme value or the derivative is plotted.

■ 3.4 Example: The Cosine Function

To illustrate the use of the Taylor series including bounding the error, let us approximate the cosine function, arbitrarily, through the third derivative term. Let x be 60 degrees $= \dfrac{\pi}{3}$ radians and x_0 be 45 degrees $= \dfrac{\pi}{4}$ radians. (Remember, angles must be in radians.) This example is deliberately simple to illustrate the process rather than the obvious result.

Let
$$f(x) = \cos(x)$$
$$f^1(x) = -\sin(x)$$
$$f^2(x) = -\cos(x)$$
$$f^3(x) = \sin(x)$$
$$f^4(x) = \cos(x).$$

Since $x_0 = \dfrac{\pi}{4}$, the function and derivatives at x_0 are boringly repetitive:

$$f(x_0) = \cos\left(\frac{\pi}{4}\right) = 0.707107$$

$$f^1(x_0) = -\sin\left(\frac{\pi}{4}\right) = -0.707107$$

$$f^2(x_0) = -\cos\left(\frac{\pi}{4}\right) = -0.707107$$

$$f^3(x_0) = \sin\left(\frac{\pi}{4}\right) = 0.707107$$

$$f^4(x_0) = \cos\left(\frac{\pi}{4}\right) = 0.707107$$

and

$$x - x_0 = \frac{\pi}{3} - \frac{\pi}{4} = 0.261799.$$

Substituting these in (3.3.1), we obtain

$$\cos\left(\frac{\pi}{3}\right) \approx 0.707107 - 0.707107 * 0.261799 - \frac{0.707107 * (0.261799)^2}{2!}$$
$$+ \frac{0.707107 * (0.261799)^3}{3!}$$
$$= 0.707107 - 0.185120 - 0.024232 + 0.002115$$
$$= 0.499869.$$

E$_A$ after 1 deriv. term   E$_A$ after 2 deriv. term   E$_A$ after 3 deriv. term

The correct answer is, of course, 0.5 so we are very close. Hence, the true error is as follows:

$$E_T = |0.5 - 0.499869| = 0.000131.$$

In this simple case, we easily can compute the true error because we know the correct value. The approximate error resulting from comparing the first two terms of the series is as follows:

$$E_A = |-0.185120 - 0.707107| = 0.892227.$$

Notice the sign change made the difference additive, and the error is large.

If we compare the third and second terms, we have

$$E_A = |-0.024232 - (-0.185120)| = 0.160888,$$

which is quite a bit better.

Finally, if we compare the fourth and third terms we have:

$$E_A = |0.002115 - (-0.024232)| = 0.026347.$$

Clearly, the terms are bunching, a good sign.

Let us now consider the remainder term in order to estimate the true error. In a practical situation we will not know the exact answer and therefore computing the true error will be impossible. We shall deliberately overestimate its value (i.e., bound the error). If we can't get it correct, we want our estimate to be conservative.

The remainder term after three derivatives will use the fourth derivative, which is $\cos(z)$, with z bounded by x_0 and x, i.e.,

$$\frac{\pi}{4} <= z <= \frac{\pi}{3}.$$

In the first quadrant $\left(x < 90 \text{ degrees} = \frac{\pi}{2} \text{ radians} \right)$, the cosine function decreases steadily (monotonically) from 1 to 0. Hence, its value at $x = \frac{\pi}{4}$ is greater than its value at $\frac{\pi}{3}$ (in magnitude, we do not care about sign, in general). Thus, we choose $z = \frac{\pi}{4}$ and

$$f^{IV}(z) = f\left(\frac{\pi}{4}\right) = \cos\left(\frac{\pi}{4}\right) = 0.707107.$$

So,

$$R_3 = \frac{f^{IV}(z) * (x - x_0)^4}{4!} = \frac{0.707107 * (0.261799)^4}{24} = 0.000138$$

and

$$R_3 = 0.000138 > E_T = 0.000131.$$

As desired, this is greater than (or equal to) the true error, as it always must be if we do this correctly. (Occasionally, a round-off error can give us a contrary result if the true and estimated errors are very close, but this will not happen often.)

The signs of R_n and E_T will be the same. If they are negative, we require the magnitude of R_n be greater than the magnitude of E_T: that is,

$$|R_n| > |E_T|.$$

■ 3.5 Maclaurin Series

A *Maclaurin series* is simply a Taylor series in which x_0 is 0. Thus, it is an expansion about the origin. Maclaurin series expansions for trigonometric and other common transcendental functions are often provided in the appendices of engineering and physics textbooks.

Let us return to the cosine function but this time expand it about the origin as a Maclaurin series through the third derivative term:

$$f(x_0) = \cos(0) = 1.0 \qquad f(x) = \ln(x) \qquad f(x) = e^x$$
$$f^1(x_0) = -\sin(0) = 0.0 \qquad f'(x) = \frac{1}{x} \qquad f'(x) = -e^{-x}$$
$$f^2(x_0) = -\cos(0) = -1.0 \qquad f''(x) = \frac{-1}{x^2} \qquad f''(x) = e^{-x}$$
$$f^3(x_0) = \sin(0) = 0.0. \qquad f'''(x) = \frac{2}{x^3} \qquad f'''(x) = -e^{-x}$$

Then,

$$f(x) = \cos(x) \approx \cos(x_0) - \sin(x_0) * (x - x_0)$$
$$+ \frac{1}{2!} * (-\cos(x_0)) * (x - x_0)^2 + \frac{1}{3!} * \sin(x_0) * (x - x_0)^3$$
$$= 1 - 0 * (x - 0) - \frac{1}{2!} * (x - 0)^2 + \frac{1}{3!} * 0 * (x - 0)^3$$
$$\cos(x) \approx 1 - 0.5 * x^2.$$

If we apply this expansion to $x = 60$ degrees $= \dfrac{\pi}{3}$ radians, we obtain

$$\cos\left(\frac{\pi}{3}\right) \approx 1 - 0.5 * \left(\frac{\pi}{3}\right)^2 = 0.4517,$$

with a true error of $0.5 - 0.4517 = 0.0483$. This is clearly not as good as with the Taylor series. But this is to be expected since $x - x_0$ in the Taylor series was 0.261799 while in the present Maclaurin series, $x - x_0$ is $\dfrac{\pi}{3} = 1.047$.

If we bound the error by conservatively evaluating the fourth derivative in the remainder term at $x = 0$, we have

$$R_3 = \frac{f^{IV}(z) * (x - x_0)^4}{4!} = \frac{\cos(z) * (1.047)^4}{24} = \frac{1 * 1.047^4}{24} = 0.0501.$$

This is greater than the true error, so we have properly bounded the error.
   If x is only 5 degrees = 0.0873 radians, we obtain

$$\cos(0.0873) = 1 - 0.5 * (0.0873)^2 = 0.9962,$$

which agrees with the correct value through four significant figures. On the other hand, if x = 90 degrees = 1.571 radians, the two-term series yields a value for $\cos(x)$ of -0.234, while the correct value is 0. Clearly, the distance of x from 0 is very important in a Maclaurin series.

■ 3.6 An Exponential Example

Let $f(x) = e^{(-x)}$.
   If $x_0 = 0$, that is, we are writing another Maclaurin series, $f(x_0) = e^{(-0)} = 1$.
Then,

$$f^I(x_0) = -e^{(-x_0)} = -1$$
$$f^{II}(x_0) = e^{(-x_0)} = 1.$$

Our Maclaurin series through the second derivative term (far enough) becomes

$$e^{(-x)} = 1 - x + \frac{x^2}{2} + R_2.$$ (3.6.1)

Let $x = 0.5$.
Then, $e^{(-.5)} = 0.6065$ while (3.6.1) yields

$$1 - (.5) + \frac{(0.5)^2}{2} = 0.6250.$$

The true error is $0.6065 - 0.6250 = -0.0185$.
   Note the error is negative. We will estimate this error by again conservatively evaluating the remainder term with $f^{III}(z) = -e^{(-z)}$. Clearly, this is a maximum in magnitude when $z = 0$.

Then,

$$R_2 = \frac{f^{III}(z) * (x - x_0)^3}{3!}$$

$$= -1 * \frac{x^3}{6} = -\frac{0.5^3}{6} = -0.02083.$$

We obtained the correct sign of the error and it is properly bounded, because $|R_2| = |-0.02083| > |E_T| = |-0.0185|$.

3.6.1 Using the Exponential Example

Suppose we need to find the value of x that makes the function

$$f(x) = e^{-x} - x = 0. \tag{3.6.2}$$

That is, we want the root of this equation. We shall consider root finding in detail in Chapter 7. In the meantime, we can use the Maclaurin series expansion for e^{-x} to simplify this function and find an approximate value for the root.

If e^{-x} is approximated as $1 - x + \dfrac{x^2}{2}$ as in (3.6.1), we can write the function as

$$f(x) \approx 1 - x + \frac{x^2}{2} - x = 0$$

$$\approx 1 - 2x + \frac{x^2}{2} = 0.$$

This is a quadratic equation, which is easily solved to yield roots of $x = 0.586$ and $x = 3.42$. A correct root to the function in (3.6.2) is 0.567. Thus, our approximation to the function easily yielded a reasonable estimate of a root.

If we retained only the first two terms in our Maclaurin series representation of e^{-x}, that is,

$$e^{-x} = 1 - x,$$

and substituted this in (3.6.2), we would get the simplified equation

$$f(x) \approx 1 - x - x = 0.$$

This is even easier to solve and yields $x = 0.5$, which is still not a bad first approximation.

■ 3.7 Uncertainty

In engineering and science, the numbers we use come from measurements of size and physical properties. Measurements are never exact; therefore, even the input to our calculations contain some uncertainty. Let us now briefly consider how uncertainty in input can affect the results we obtain, even if our calculations are mathematically correct.

Consider the following calculation:

$$u = x * (y + z),$$

where the values of x, y, and z contain uncertainty.

Let $x = \bar{x} \pm e_x$, $y = \bar{y} \pm e_y$, and $z = \bar{z} \pm e_z$, where the barred quantities are the mean values, and e_x, e_y, and e_z are the uncertainties in x, y, and z respectively. In general, the uncertainties can be plus or minus. We want to determine the uncertainty in u, e_u.

3.7.1 Exact Analysis

The uncertainty in u will be either the difference between u_{max} and $\bar{u}$ or between $\bar{u}$ and u_{min}. Since both are possible and these may not be equal, we shall simply average these two uncertainties, that is,

$$e_{u,\text{average}} = \Delta u = \frac{(u_{max} - \bar{u}) + (\bar{u} - u_{min})}{2}.$$

We shall refer to this simply as e_u.

Since the $\bar{u}$'s cancel, this becomes

$$e_u = \Delta u = \frac{(u_{max} - u_{min})}{2}. \tag{3.7.1}$$

This is the so-called exact analysis.

Given $x = 1.0 \pm 0.1$, $y = 2.0 \pm 0.2$, and $z = 5.0 \pm 0.4$, we see that $\bar{x} = 1$ and $e_x = 0.1$, and so on. If we express these uncertainties in relative form, we have

$$\frac{e_x}{\bar{x}} = \frac{0.1}{1.0} * 100 = 10\%, \qquad \frac{e_y}{\bar{y}} = \frac{0.2}{2.0} * 100 = 10\%, \quad \text{and}$$

$$\frac{e_z}{\bar{z}} = \frac{0.4}{5.0} * 100 = 8\%.$$

So the relative uncertainty in x is 10%, in y is 10%, and in z is 8%.

First, let us calculate the mean value of u, $\bar{u}$, using the mean values of x, y, and z. Then,

$$\bar{u} = \bar{x} * (\bar{y} + \bar{z})$$
$$\bar{u} = 1 * (2 + 5) = 7.$$

The uncertainty u_{max} occurs when we use the maximum values of x, y, and z. Shortly we will see that it sometimes requires care to make the correct choice of values to maximize or minimize the result. Accordingly,

$$u_{max} = (x + x_{max}) * (y + y_{max} + z + z_{max})$$
$$= (1 + 0.1) * (2 + 0.2 + 5 + 0.4) = 1.1 * (2.2 + 5.4) = 8.36.$$

Likewise,

$$u_{min} = (x - x_{max}) * (y - y_{max} + z - z_{max})$$
$$= (1 - 0.1) * (2 - 0.2 + 5 - 0.4) = 0.9 * (1.8 + 4.6) = 5.76.$$

Then, from (3.7.1), we obtain

$$e_u = \Delta u = \frac{(8.36 - 5.76)}{2} = 1.30.$$

This is the exact uncertainty. We have used the extreme values of u to calculate this result.

We may write

$$u = \bar{u} \pm e_u = 7.0 \pm 1.3$$

or

$$u_{min} <= \bar{u} <= u_{max}$$
$$7.0 - 1.3 <= \bar{u} <= 7.0 + 1.3$$
$$5.7 <= \bar{u} <= 8.3.$$

We know that u is somewhere in this range and we cannot say with greater certainty than this, given the uncertainty in our input data.

Another way to represent this uncertainty is to compare e_u to $\bar{u}$. We are in for a surprise, because the relative uncertainty in e_u is

$$\frac{e_u}{\bar{u}} = \frac{\Delta u}{\bar{u}} = \frac{1.3}{7} * 100 = 19\%.$$

So the three variables, x, y, and z, with a maximum uncertainty in any of 10%, have been combined in one calculation to produce an answer with an uncertainty of 19%. The uncertainty has grown significantly in one simple calculation! Thus, we must do all we can to reduce additional uncertainty caused by round-off or truncation errors in our calculations.

3.7.2 Approximate Analysis

An alternate way of estimating uncertainty is to perform an approximate analysis. This has the advantage that it can identify the source of the greatest uncertainty, and it also may be easier to do.

Again, let us consider the calculation $u = x * (y + z)$ in which the values of x, y, and z contain uncertainty.

Let each term be expressed in terms of its mean value and its uncertainty; for example, $x = \bar{x} \pm e_x$, $y = \bar{y} \pm e_y$, and $z = \bar{z} \pm e_z$. Substituting in the expression for u, we obtain

$$\bar{u} \pm e_u = (\bar{x} \pm e_x) * (\bar{y} \pm e_y + \bar{z} \pm e_z).$$

Expanding the expression yields

$$\bar{u} \pm e_u = \bar{x} * \bar{y} \pm \bar{x} * e_y + \bar{x} * \bar{z} \pm \bar{x} * e_z \pm e_x * \bar{y} \pm e_x * e_y \pm e_x * \bar{z} \pm e_x * e_z$$
$$= \bar{x} * \bar{y} + \bar{x} * \bar{z} \pm \bar{x} * e_y \pm \bar{x} * e_z \pm e_x * \bar{y} \pm e_x * \bar{z} \pm e_x * e_y \pm e_x * e_z.$$

The products of the uncertainties ($e_x * e_y$ and $e_x * e_z$) are assumed to be small and can be neglected with respect to the other terms in the equation. Thus, we linearize the equation, leaving

$$\bar{u} \pm e_u = \bar{x} * \bar{y} + \bar{x} * \bar{z} \pm \bar{x} * e_y \pm \bar{x} * e_z \pm e_x * \bar{y} \pm e_x * \bar{z}.$$

Collecting terms, we have

$$\bar{u} \pm e_u = \bar{x} * (\bar{y} + \bar{z}) \pm \bar{x} * (e_y + e_z) \pm e_x * (\bar{y} + \bar{z}).$$

But $\bar{u} = \bar{x} * (\bar{y} + \bar{z})$. If we subtract $\bar{u}$ from the left side of the equation and its equivalent $\bar{x} * (\bar{y} + \bar{z})$ from the right side, we obtain

$$e_u = \pm \bar{x} * (e_y + e_z) \pm e_x * (\bar{y} + \bar{z}).$$

If we replace the uncertainties by differentials (also small quantities), we have

$$du = \pm \bar{x} * (dy + dz) \pm dx * (\bar{y} + \bar{z}).$$

But this is just the differential of u, following the usual method for the differentiation of a product. Hence, the expression for the uncertainty of a quantity is the same as its differential, which makes it easy to remember. Applied to our problem and substituting the uncertainties for the differentials, we obtain

$$e_u = du = \bar{x} * (e_y + e_z) + e_x * (\bar{y} + \bar{z})$$
$$e_u = du = 1 * (0.2 + 0.4) + 0.1 * (2 + 5)$$
$$e_u = du = 0.6 + 0.7 = 1.3.$$

This is the same result as before. Usually, the two methods will not yield identical results, but in this case the uncertainties occurred in linear fashion in the equation, and therefore a linear approximation yielded an exact result. We only considered the plus sign where $\pm$ appeared in the expression because we are looking for the maximum uncertainty. Hence, we have to guard against cancellation of positive and negative terms. In the following, we will use an alternate form that takes absolute values for the same reason.

A slightly different but more useful form of approximate analysis would be to sum the magnitudes of the partial derivatives:

$$e_u = du = \left| \frac{\partial u}{\partial x} e_x \right| + \left| \frac{\partial u}{\partial y} e_y \right| + \left| \frac{\partial u}{\partial z} e_z \right|, \qquad (3.7.2)$$

where each term accounts for the uncertainty caused by the uncertainty in a given input variable (Chapra and Canale, 2005). We use the absolute value of the individual contributions because we must assume the worst case, which occurs when all uncertainties add together. This might not actually happen, but we cannot assume otherwise. In working a problem, it is possible to make two equal mistakes of opposite signs and still get the right answer. But it cannot be assumed that the signs of the uncertainties will offset one another. Now, let us consider that

$$\frac{\partial u}{\partial x} = y + z = 2 + 5 = 7$$
$$\frac{\partial u}{\partial y} = x = 1 \text{ and } \frac{\partial u}{\partial z} = x = 1.$$

Then,

$$e_u = du = |7 * (0.1)| + |1 * (0.2)| + |1 * (0.4)|$$
$$e_u = du = 0.7 + 0.2 + 0.4 = 1.3$$

as before, and 0.7 is the uncertainty in u caused by the uncertainty in x. Since it is greater than the uncertainties caused by y and z (0.2 and 0.4), we know that x is the largest source of uncertainty. If we could obtain a more accurate value of x, our result would improve.

For example, suppose the uncertainty in x is reduced to 0.05. Then,

$$e_u = du = 7 * (0.05) + 1 * (0.2) + 1 * (0.4)$$
$$e_u = du = 0.35 + 0.2 + 0.4 = 0.95.$$

The uncertainty in u has been reduced by 0.35, and z has become the greatest contributor to the overall uncertainty. By way of analogy, if the uncertainties in x, y, and z came from manufacturing tolerances, the uncertainty in u would be the variation in u resulting from these tolerances. By using the partial derivative chain, we could identify which input tolerance was the greatest source of the uncertainty in our result and perhaps make the appropriate changes to improve the manufacturing process to improve our product.

3.7.3 Summary

We have considered exact and approximate analyses, and absolute and relative uncertainty. Thus, we have two procedures and two forms to express the uncertainty: a total of four combinations. These are summarized in Table 3.1.

TABLE 3.1 Uncertainty Analysis Types and Forms

Analysis type	Absolute uncertainty	Relative uncertainty
Exact analysis	Δu	$\dfrac{\Delta u}{\bar{u}}$
Approximate analysis	du	$\dfrac{du}{\bar{u}}$

■ 3.8 A More Complicated Example

Let $w = \dfrac{y - x}{z}$ and again use $x = 1.0 \pm 0.1$, $y = 2.0 \pm 0.2$, and $z = 5.0 \pm 0.4$.

Then, $\overline{w} = \dfrac{2.0 - 1.0}{5.0} = 0.2$.

If we perform an exact analysis of the uncertainty, we obtain

$$w_{\max} = \frac{y_{\max} - x_{\min}}{z_{\min}} = \frac{2 + 0.2 - (1 - 0.1)}{5 - 0.4} = \frac{2.2 - 0.9}{4.6} = 0.283.$$

Notice that in maximizing w, we had to take the maximum numerator and the minimum denominator. Taking the greatest possible spread between y and x—that is, using $y_{\max}$ and $x_{\min}$—maximizes the numerator. Getting this right requires careful thinking. Thus,

$$w_{\min} = \frac{y_{\min} - x_{\max}}{z_{\max}} = \frac{2 - 0.2 - (1 + 0.1)}{5 + 0.4} = 0.130.$$

We had to minimize the numerator and maximize the denominator to obtain $w_{\min}$. Then,

$$e_w = \Delta w = \frac{(w_{\max} - w_{\min})}{2} = \frac{0.283 - 0.130}{2} = 0.0765.$$

The relative uncertainty in w is

$$\frac{\Delta w}{\overline{w}} = \frac{0.00765}{0.2} * 100 = 38\%,$$

a big increase from the uncertainty in the input variables!

If we perform an approximate analysis to determine the uncertainty, we obtain

$$e_w = dw = \left| \frac{\partial w}{\partial x} e_x \right| + \left| \frac{\partial w}{\partial y} e_y \right| + \left| \frac{\partial w}{\partial z} e_z \right|.$$

Now

$$\frac{\partial w}{\partial x} = -\frac{1}{z} = -\frac{1}{5} = -0.2$$

and

$$\frac{\partial w}{\partial y} = \frac{1}{z} = 0.2 \quad \frac{\partial w}{\partial z} = \frac{-(y-x)}{z^2} = \frac{-(2-1)}{5^2} = -0.04.$$

Then,

$$e_w = dw = |-0.2 * 0.1| + |0.2 * 0.2| + |-0.04 * 0.4|$$

$$= 0.02 + 0.04 + 0.016 = 0.076.$$

The relative uncertainty is

$$\frac{e_w}{w} = \frac{dw}{w} = \frac{0.076}{0.2} * 100 = 38\%$$

as before.

3.8.1 A Travel Example

Consider a trip by air from Los Angeles to New York City. The distance traveled and the average speed will vary because of winds and weather. Assume the distance is 2750 ± 250 miles and the groundspeed (actual speed over the ground) is 550 ± 50 mph. Accordingly, $V = \frac{d}{t}$ where V is the velocity, d is the distance, and t is the time. Solving for t, we have, $t = \frac{d}{V}$. And,

$$\bar{t} = \frac{\bar{d}}{\bar{V}} = \frac{2,750}{550} = 5.0 \text{ hrs.}$$

What is the uncertainty in the flight time? In performing an exact analysis, we see that

$$t_{max} = \frac{d_{max}}{V_{min}} = \frac{2750 + 250}{550 - 50} = 6.0 \text{ hrs.}$$

Thus, if the plane has to fly the longest route at the slowest groundspeed, we get the maximum flight time of 6 hours. On the other hand, if the plane can fly the shortest route at the highest groundspeed we get the minimum flight time:

$$t_{min} = \frac{d_{min}}{V_{max}} = \frac{2750 - 250}{550 + 50} = 4.2 \text{ hrs.}$$

The expected uncertainty will be in between:

$$e_t = \Delta t = \frac{6 - 4.2}{2} = 0.9 \text{ hrs,}$$

an uncertainty of almost an hour!

Performing an approximate analysis yields

$$\frac{\partial t}{\partial V} = \frac{-d}{V^2} = \frac{-2750}{550^2} = -0.0091 \text{ hr}^2/\text{mi}$$

and $\dfrac{\partial t}{\partial d} = \dfrac{1}{V} = \dfrac{1}{550} = 0.0018 \text{ hr/mi.}$

Then $e_t = dt = \left| \dfrac{\partial t}{\partial V} e_V \right| + \left| \dfrac{\partial t}{\partial d} e_d \right|$, and

$$e_t = dt = |0.0091 * 50| + |0.0018 * 250| = 0.46 + 0.45 = 0.91 \text{ hrs.}$$

This result is almost the same as our exact analysis. It shows us that, in this case, uncertainty in distance and velocity contribute almost equally to the uncertainty in the result.

The airlines include reasonable allowances in their schedules for this uncertainty. On long flights like this one, it is easy to make up for a late start, but that is not possible on short ones.

3.8.2 An Example for Later

Consider the fraction $m = \dfrac{N}{D}$, where N and D are both integers. We may express the uncertainty in m as

$$e_m = \left| \frac{\partial m}{\partial N} u_N \right| + \left| \frac{\partial m}{\partial D} u_D \right|$$

$$\frac{\partial m}{\partial N} = \frac{1}{D} \quad \text{and} \quad \frac{\partial m}{\partial D} = -\frac{N}{D^2}.$$

If $D > N$, both derivatives will be less than 1 in magnitude, minimizing the uncertainty in m. We shall return to this point in Chapter 9 when we consider partial pivoting in solving sets of linear equations.

■ 3.9 Chapter 3 Exercises

3.1 Write a Maclaurin series expansion for $\sin(x)$ through the third derivative term. Use your series to estimate $\sin(20°)$. Bound your error and compare to the exact answer.

3.2 Write a Maclaurin series expansion for $\sec(x)$ through the third derivative term. Use your series to estimate $\sec(20°)$. Bound your error and compare to the exact answer.

3.3 Consider the equation $f(x) = \sec(x) - \cos(x) = 0.7$, x in radians. Use your general Maclaurin series for $\sec(x)$ from Exercise 3.2. Approximate $\cos(x)$ as $1 - \dfrac{x^2}{2}$. Substitute these approximations in the given equation and solve for an approximate value of x.

3.4 Write a Taylor series for $\ln(x)$ through the third derivative term. Expand about $x_0 = 1$. Apply your series to $x = 1.5$. Bound your error and compare to the exact answer.

3.5 Write a Maclaurin series for $f(x) = e^x$ through the third derivative term. Apply your series to estimating e^2. Bound the error of your series and compare to the true error.

3.6 Write a Maclaurin series expansion for $\sin^{-1}(x)$ through the third derivative term. Use your series to estimate $\sin^{-1}(0.5)$. Bound your error and compare to the exact answer.

3.7 Consider the equation $f(x) = e^{(-x)} - 2 * \cos(x) = 0$, x in radians. Express both $e^{(-x)}$ and $\cos(x)$ as a Maclaurin series. Exercise 3.5 may be helpful, but note the sign has changed on the exponential. Retain only terms through x^2. Substitute these approximations into the given equation and solve for an approximate value of x.

3.8 Write a Taylor series expansion for $\tan(x)$ through the third derivative term. Use your series to estimate $\tan(70°)$ by expanding about $x_0 = 45°$. Bound your error and compare to the exact answer.

3.9 Write a Maclaurin series for $\sinh(x)$ through the third derivative term. Note $\dfrac{d\,\sinh(x)}{dx} = \cosh(x)$ and $\dfrac{d\,\cosh(x)}{dx} = \sinh(x)$ (no sign change).

3.10 Let $f(x) = x^{1/3}$

    a. Develop a Taylor series through the third derivative term for this function.
    b. Use your series to estimate the cube root of 300 by expanding about $x_0 = 343$.
    c. Bound the error of your estimate and compare it to the true error.

3.11 A certain piston has a diameter of 0.875 ± 0.001 in. It travels within a cylinder of diameter 0.881 ± 0.001 in. What is the uncertainty in the clearance between piston and cylinder?

3.12 Newton's law of universal gravitation is

$$F = \frac{Gm_1m_2}{r^2}.$$

G can be determined if F, m_1, m_2, and r are measured for two objects. Then

$$G = \frac{Fr^2}{m_1m_2}.$$

Given that $F = 13 \times 10^{-11}$ N, $\pm 2\%$, $m_1 = 0.8 \pm 0.005$ kg, $m_2 = 0.004 \pm 0.001$ kg, and $r = 0.04 \pm 0.001$ m, determine the uncertainty in G. Perform both exact and approximate analyses and present your answers in absolute and relative form. Which input variable contributes the most to this uncertainty?

3.13 For the motion of a body in a circular orbit around a central body, the radius of the orbit, r, may be related to the orbiting body's mass, m, the universal gravitational constant, G, and its period of revolution T by the equation:

$$T^2 = \frac{4\pi^2r^3}{Gm}.$$

If $G = 6.67 \times 10^{-11} \pm 0.01$ N $-$ m^2/kg^2, and for the Earth-moon system, $r = 3.84 \times 10^8$ m $\pm 2\%$ and $m = 5.97 \pm 0.20 \times 10^{24}$ kg, calculate the mean value of the moon's period, and the absolute and relative uncertainty of the period. Perform both exact and approximate analyses and present your answers in absolute and relative form. Use time units of days for your answers and retain three significant figures.

3.14 The velocity of a flow may be measured using a manometer, a pitot-static tube, and the following formula:

$$dV = \left| \frac{\partial V}{\partial \gamma} e_\gamma \right| + \left| \frac{\partial V}{\partial h} e_h \right| + \left| \frac{\partial V}{\partial \rho} e_\rho \right|$$

$$dV = \left(\frac{2h}{\rho} \right)^{1/2} \left(\frac{1}{2} \right) \gamma^{-1/2}$$

$$V = \sqrt{\frac{2 * \gamma * h}{\rho}}$$

where γ is the specific weight of the manometer fluid, h is the differential height in the manometer legs, and ρ is the density of the flowing fluid.

Given $\gamma = 57.0 \pm 0.15$ lb/ft^3, $h = 0.15 \pm 0.01$ ft, and $\rho = 0.00238 \pm 0.0001$ slug/ft^3, determine the speed of the flow and its uncertainty. Perform both exact and approximate analyses and present your answers in absolute and relative form.

3.15 The equation of state for an ideal gas is

$$PV = mRT.$$

For air, $R = 53.3$ ft-lb/(lbm-°R). Given m $= 1$ lbm, $P = 2100 \pm 0.2$ lb/ft^2, and $T = 500 \pm 0.5$°R. What is the uncertainty in V? Perform both exact and approximate analyses and present your answers in absolute and relative form.

4 Decisions and Loops: Which Is Bigger? How Many Times?

■ 4.1 Why Comparisons?

Computer programs are typically written to be run with a general range of input values. For example, if a program segment required the solution of a quadratic equation of the form

$$ax^2 + bx + c = 0,$$

the roots may be real or complex, depending on the value of the discriminant

$$d = b^\wedge 2 - 4 * a * c.$$

VBA is not capable of taking the square root of a negative number. Therefore, a program to solve a quadratic equation would have to recognize if d were negative before proceeding. Attempting to evaluate the square root of a negative d would produce a run-time error message,

```
Invalid procedure call or argument
```

Other decisions might involve using different material properties in different temperature regimes such as the specific heat of a gas,

$$Cp = a_1 + b_1 * T + c_1 * T^2 \text{ for } T < \, = Tr,$$

but

$$Cp = a_2 + b_2 * T + c_2 * T^2 \text{ for } T > Tr.$$

Here, $a_1, a_2, b_1, b_2, c_1,$ and c_2 are constants, and T is the current temperature. Hence, T would need to be compared to a reference temperature Tr before the proper value of Cp could be determined.

Another example is the formula for the work, W, in a polytropic thermodynamic process

$$W = \frac{P_2V_2 - P_1V_1}{1 - n}, n \neq 1$$

but $W = (p_1V_1) * \ln\left(\frac{V_2}{V_1}\right)$ for $n = 1$.

Here the P's are pressures, the V's are volumes, and n is the polytropic exponent that defines the process. Thus, the value of n would have to be compared to 1 before the proper formula could be selected.

■ 4.2 If Statements

Decisions in VBA are often made with If statements. The general form is

```
If condition Then
     . . . . .
     . . . . .
End If
```

where condition is a statement that is either true or false. The programming segment begins with the If statement and ends with an End If statement. If the condition is true, the statements following If up to End If are executed. If the condition is false, these statements are skipped and the program proceeds to the line following End If.

For example, in a program to find the roots of a quadratic equation, we might have

```
d = b^2 - 4*a*c
If d >= 0 then
     r1 = (-b + sqr(d))/(2*a)
     r2 = (-b - sqr(d))/(2*a)
End If
```

Notice that we have used a compound condition in the If statement. The statements following it are performed if either $d = 0$ or d is > 0.

TIP

We have indented the statements between the If and End If lines to emphasize that they are contingent. This is not necessary, but doing so makes the program easier to read.

In the quadratic equation problem, d may be less than 0. If this occurs, there is a complex solution involving real and imaginary parts. We can consider both real and complex outcomes using an Else statement with our If statement as follows:

```
d = b^2 - 4*a*c
If d >= 0 Then
     r1 = (-b + sqr(d))/(2*a)
     r2 = (-b - sqr(d ))/(2*a)
     Msgbox "r1 = " & r1 & " r2 = " & r2
Else
     r = -b/(2*a)
     cpx = sqr(-d)/(2*a)
     Msgbox "r1 = " & r & cpx & " i"
     Msgbox "r2 = " & r & -cpx & " i"
End If
```

If d is not equal to or greater than zero, the program skips the lines between the If line and the Else line and instead executes the four lines following the Else. We have included the output within the If construction because the form of the output also changes depending on whether the roots are real or complex. A flowchart for a program to solve a general quadratic equation is in Figure 4.1.

■ 4.3 ElseIf

If more than two choices are possible for the initial test, the ElseIf statement should be used. The general structure is as follows:

```
If condition 1 Then
     .....
     .....
ElseIf condition 2 Then
     .....
     .....
ElseIf condition 3 Then
     .....
     .....
     ....
     etc.
Else
     .....
     .....
End If
```

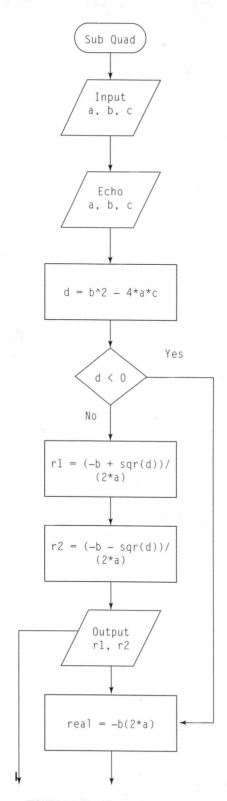
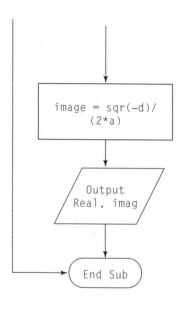

FIGURE 4.1 Quad Program Flowchart.

For example, if we wished to check a class schedule we could write the following program segment. Since strings are case-sensitive, `Monday` and `monday` would be interpreted differently by VBA.

```
Dim today as string
today = InputBox("Enter day name, use initial capital,
e.g. Tuesday")
If  today = "Monday" Or today = "Wednesday" Then
    MsgBox " ENGR 201 10 a.m. " & "ENGR 221 1 p.m."
ElseIf today = "Tuesday" Or today = "Thursday" Then
    MsgBox " ENGR 231 8 a.m. " & "ENGL 201 10:50 a.m."
    & "ENGR 255 1:40 p.m."
ElseIf today = "Friday" Then
    MsgBox " ENGR 202 1 p.m."
Else
    MsgBox "No classes"
End If
```

■ 4.4 Boolean Operations

So far we have used only the = (equal) and > (greater than) symbols in comparisons. The list of VBA Boolean (comparison) operators includes these:

=	equal	x = y
>	greater than	x > y
<	less than	x < y
>=	greater than or equal	x >= y
<=	less than or equal	x <= y
<>	not equal	x <> y (both greater than and less than symbols together)

In addition, symbols may be combined with `And` or `Or` and grouped within parentheses. We might have written the test for a positive root less compactly as

```
If d = 0 Or d > 0 Then
    . . . . .
    . . . . .
End If
```

The statements following the `If` would be performed if either d were equal to 0 or d were greater than 0. But

```
If d = 0 And d > 0 Then
    . . . . . .
    . . . . . .
End If
```

would produce no result, because *d* cannot simultaneously be equal to 0 and greater than 0. Statements using And are performed only if the combined conditions are true, and not performed otherwise.

Careful logic is required in writing the conditions for If statements. Suppose we are trying to find the roots of a pair of nonlinear equations for roots *x* and *y* by iteration. (We will do this in Chapter 7.) Each iteration produces an approximate error in *x*, perhaps named errx, and an approximate error in *y* named erry. We want the iteration to continue as long as either errx or erry is greater than some tolerance, tol. But we do not want the process to run on forever if it does not converge, and therefore we also will specify a maximum number of iterations, itmax.

We might control this with:

```
If iter < itmax And (errx > tol Or erry > tol) Then
      . . . . .
      . . . . .
End If
```

The process will continue as long as both the first condition (iter < itmax) and either part of the second condition (errx > tol) or the (erry > tol) is true.

■ 4.5 Sine Series Program

A Maclaurin series for the sine of an angle, *x*, *x* in radians, is given by:

$$\sin(x) = x - \frac{x^3}{3!} + \frac{x^5}{5!} - \frac{x^7}{7!} + \ \ldots\ldots$$

The series uses only odd powers of *x*, and the signs of the terms alternate.

We could begin a program to evaluate this series by entering the value of *x*, say in degrees, the tolerance to be applied to the last term, and the maximum number of terms. We must convert *x* to radians. We then initialize the sum to *x* (the first term in this series), define the first term as *x*, the number of terms computed, *n*, as 1, the power, *p*, as 1, and the factorial, *fact*, as 1: that is,

```
. . . . .
p = 1
n = 1
fact = 1
term = x
sign = 1
sum = x
. . . . .
```

To add additional terms until either we encountered a term less, in absolute value, than our tolerance or reached our limit for the number of allowed terms, we could continue our sine program as follows:

```
1:  p = p + 2
n = n + 1
fact = fact*p*(p-1)
term = x^p/fact
sign = -sign
sum = sum + sign*term
If n < maxterms And abs(term) > tol Then
     goto 1
End If
.....
```

Several things should be noticed. First, instead of computing a complete factorial, we have recognized that, each time, the new factorial is just the old factorial multiplied by the new power, p, and the new power minus 1, that is, $p - 1$. Doing this saves time compared to completely evaluating the factorial for each term. Second, we have accounted for the sign change by each time reversing the value of the previous sign by multiplying the sign by -1. Finally, we compared the absolute value of the current term with tol and the number of the current term with the maximum number of allowed terms in a compound If statement. We needed to use the absolute value of term because we are interested only in its magnitude. Any negative value of term would be less than a positive tol. If both conditions for continuation are satisfied, the line goto 1 sends the program back to the line designated 1 where the power is increased by 2, following which the next term is generated.

Thus, we have created a loop to evaluate the terms of a sine series until either we have calculated a term sufficiently small that we choose not to compute another one, or we have calculated the maximum number of terms. It is good programming practice to limit the maximum number of terms in case the series does not converge, whether because we have made a programming mistake or we are dealing with a truly diverging series. Without such a limit, the computer would happily continue generating more and more terms for a long while.

■ 4.6 Conditional Loops: The Do While Loop

The loop we have just created is a conditional loop. We do not know in advance how many times the program will go through the steps of the loop. It depends on the values of n and term. Such loops occur often in calculations, and hence high-level programming languages like VBA have a special procedure that provides for both testing and looping in such calculations. This eliminates the need to number statements and include `goto` statements whose presence can make programs hard to read. Using line numbers and `goto`'s produce what is often referred to as spaghetti code. The procedures that follow are called structural programming and are much preferred.

In VBA, the `Do While` statement controls conditional loops. The general form of the `Do While` statement is

 # Continues as long as Condition is true.

```
Do While condition
     . . . . .
     . . . . .
     . . . . .
Loop
```

The loop continues to be performed as long as the condition is true. The condition to be checked is written in the same form as in an `If` statement. When it becomes false, program execution skips to the line immediately after `Loop`.

The loop in the previous sine program, beginning with line 1: `p = p + 2` could be written more simply using `Do While`, as follows:

```
Do While n < maxterms And abs(term) > tol
     p = p + 2
     n = n + 1
     fact = fact*p*(p-1)
     term = x^p/fact
     sign = -sign
     sum = sum + sign*term
Loop
```

A flowchart for the sine series program is shown in Figure 4.2.

4.6.1 The Square Root of a Number

The square root of a number can be computed by iterating on the equation

$$xn = \frac{\left(x + \dfrac{A}{x} \right)}{2},$$

where A is the number whose square root we seek, x is the current estimate, and xn is the next guess. For example, if $A = 60$ and our initial guess for $\sqrt{60}$ is 8 (reasonable since $\sqrt{64} = 8$), the first iteration produces

$$xn = \frac{\left(8 + \dfrac{60}{8} \right)}{2} = 7.75.$$

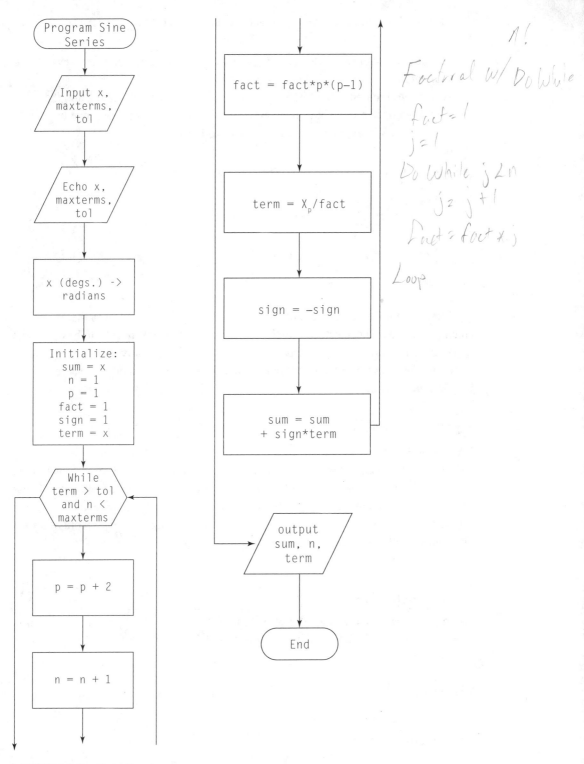

Program Sine
Series

Input x,
maxterms,
tol

Echo x,
maxterms,
tol

x (degs.) ->
radians

Initialize:
sum = x
n = 1
p = 1
fact = 1
sign = 1
term = x

While
term > tol
and n <
maxterms

p = p + 2

n = n + 1

fact = fact*p*(p−1)

term = X_p/fact

sign = −sign

sum = sum
+ sign*term

output
sum, n,
term

End

$n!$

Factral w/ Do While
fact=1
j=1
Do While j < n
 j = j + 1
 fact = fact * j
Loop

I FIGURE 4.2 Sine Series Flowchart.

The next value of *xn* would use 7.75 for *x*, and so forth. Each time, the new *x* is the just-computed *xn*. Convergence is generally rapid.

Assume *A*, the first *x*, the tolerance on the accuracy of the root, `tol`, and the maximum number of iterations have been already specified. The iterative procedure could be continued in **VBA** with a `Do While` loop as follows:

```
it = 1
diff = 1.1*tol
Do While it < itmax And diff > tol
     xn = (x + A/x)/2
     diff = abs(xn - x)
     x = xn
     it = it + 1
Loop
```

Notice we had to define an initial value for `diff` prior to the test in the `Do While` condition. If we had neglected to do this, VBA would have taken `diff` to be 0, and the loop would not have been performed at all. Setting `diff` initially to `1.1*tol` is arbitrary but convenient. The important point is that it be greater than `tol` to ensure that the loop is executed at least once.

4.6.2 Reserved Words

We have to be careful in naming variables in VBA. In the square root routine just considered, we might have been tempted to name the absolute value of `xn - x` as `Error`. But `Error` is a reserved word in VBA and must not be used as a variable name. Other reserved words we have already encountered include `Abs`, `And`, `Do`, `Or`, and `While`. Less familiar reserved words include `Len`, `New`, and `Scale`. Using a reserved word as a variable name would produce an error message that would not make much sense.

■ 4.7 Definite Loops: The `For` Loop

Often, we have to perform a calculation where the number of repetitions is known in advance. In such calculations the `For` loop is used. The general form of this loop, with `index` a counter, is

```
For index = first value   To   last value   Step increment
     . . . . .
     . . . . .
Next index
```

If the index is j and we wish to use j values from 1 to 10, we could have this:

```
For j = 1 To 10
     . . . . .
     . . . . .
Next j
```

In executing a For loop, the computer keeps changing the value of the index by the increment until it is outside the range between the first and last values of the index. For example, if the last value of the index, j, is 10 and the increment is 1, when j reaches 11 no more loop steps are performed, and the program drops to the line below Next j. The loop is finished. If the increment is 1—that is, we are counting by 1—the increment need not be specified. Thus, step = 1 is the default value.

If we wish to include only odd values of j, beginning with 1 and ending with 11, we could have this:

```
For j = 1 To 11 Step 2
     . . . . .
     . . . . .
Next j
```

Then j would be 1, 3, 5, 7, 9, and 11. Step identifies the number following as the increment.

We can also decrease the value of the index. This can be very convenient. If the first value of j = 11 and we want to count backward to 1, we could use this:

```
For j = 11 To 1 Step -1
     . . . . .
     . . . . .
Next j
```

Suppose we need to compute the factorial of a known variable *n* at some point in a program. Using a For loop, we can easily do this as follows:

```
fact = 1
For j = 2 To n
     fact = fact*j
Next j
```

If we had neglected to initialize fact to 1, VBA would have taken it to be 0 so we would have just multiplied 0 by all the numbers from 2 to *n*. The result would still be 0.

> **TIP**
>
> Because the index value advances beyond the last value by one increment when it is last tested, it is unwise to use it in subsequent calculations. Instead, beyond the factorial loop we should use *n*.

4.7.1 Factorial with `Do While`

A factorial could be computed using a `Do While` loop though it is not as efficient. Consider the following:

```
fact = 1
j = 1
Do While j < n
     j = j + 1
     fact = fact*j
Loop
```

To use this loop, we had to initialize the counter and increment it. These are automatically handled in the code for a `For` loop. That procedure is preferred for definite loops such as this one.

■ 4.8 Nested Loops

Loops may be placed inside loops. For example, in evaluating the sine series we could have placed a factorial `For` loop within the `Do While` loop. (It is common programming practice to indent nested loops to clearly indicate their hierarchy.) Thus:

```
Do While n < maxterms And abs(term) > tol
     p = p + 2
     n = n + 1
     fact = 1
     For i = 2 To p
          fact = fact*p
     Next p
     term = x^p/fact
     sign = -sign
     sum = sum + sign*term
Loop
```

This was not needed previously in Section 4.5 because we were able to build up our factorial as we developed terms in the sine series. But it would work. The important thing to note is that a loop begun inside another loop must end inside that loop. For example, the Next p marking the end of the For loop could not have been below Loop in the preceding program segment. Doing so would generate an error message: "Invalid Next control variable reference".

Suppose we were manipulating rows and columns in a nonsquare matrix. We might have an outer row loop and an inner column loop as follows:

```
For row = 1 to n
    For col = 1 to m
    . . . . .
    . . . . .
    . . . . .
    Next col
Next row
```

0.946083

The column loop defined by For col = 1 to m must be contained within the loop defined by For row = 1 to n. We shall return to manipulating rows and columns this way in Chapter 10.

4.8.1 For **Loops and Spreadsheet Cells**

Output to spreadsheet cells can use variable indices. Suppose we wanted to generate a table of factorial values for, say, 1 to 30 and output the results to a spreadsheet. We could use the For index in specifying the spreadsheet row or column. Consider the following program segment.

```
cells(1,1) = " i "
cells(1,2) = "factorial"
fact = 1
For n = 1 To 30
    fact = 1
    For i = 1 To n
        fact = fact*n
    Next i
cells(1 + n,1).value = n
cells(1 + n,2).value = fact
Next n
```

Notice the column headings for the integer and its factorial are put in row 1, columns 1 and 2, respectively. Then each n and its corresponding factorial are placed in subsequent rows.

Columns could be similarly identified by For indices. In all cases, care must be exercised to get proper placement without overwriting other values.

■ 4.9 Constant Calculations and Loops

Sometimes one or more operations apply to terms in the loop but are independent of the loop value. For example, suppose each term in the loop has to be multiplied by $\dfrac{2}{\sqrt{\pi}}$. Keep such constant operations out of the loop as indicated here:

```
pi = 4*atn(1)
a = 2/sqr(pi)
Sum = 0
Do While n < nmax
      term = ....
      sum = sum + term
Loop
Sum = sum*a
```

There is no need to perform this multiplication repeatedly within the loop when it can be factored out and applied only once.

■ 4.10 Chapter 4 Exercises

4.1 The area of any triangle may be calculated in terms of the lengths of its three sides using the following formulas. In the following, a, b, and c are the side lengths:

$$s = 0.5 * (a + b + c)$$
$$Area = \sqrt{[s * (s - a) * (s - b) * (s - c)]}.$$

Write a program to input a, b, and c, calculate the area, and output the input and the result. Include program steps to check that s is greater than a, b, and c. Otherwise, the term under the square root would be negative. Skip the calculation and display a "Message Box" warning if s fails this test. Apply your program to $a = 15$, $b = 27$, and $c = 35$.

4.2 Federal income tax marginal rates on taxable income for single individuals in the United States are:

Taxable income	rate
$1 to $7000	10%
$7001 to $28,400	15%
$28,401 to $68,000	25%
$68,001 to $143,300	28%
$143,301 to $311,950	33%
above $311,950	35%

Marginal means the rate applies to the increment in the given range. Thus, for $10,000, the first $7000 would be taxed at 10%, and the remaining $3000 (equal to 10,000 minus 7000) would be taxed at 15%. Thus, the tax on $10,000 would be

$$0.1 * 7000 + .15 * (10000 - 7000) = 700 + 450 = 1150.$$

Write a program to input taxable income and compute the tax owed. Apply your program to $25,000, $50,000, $75,000, and $175,000.

4.3 February usually has 28 days. But years divisible by 4 are leap years in which February has 29 days. However, centennial years are not leap years unless they are also divisible by 400. Thus, 1600 and 2000 were leap years, but 1700, 1800, and 1900 were not. Write a program to enter a year and determine if February has 28 or 29 days. Consider 2004, 2009, and 2100.

4.4 Evaluate the Maclaurin sine series using a `Do While` loop. Limit your series to 30 terms and a tolerance of $0.5e^{-6}$. Apply your series to $x = 28°$.

4.5 The general Maclaurin cosine series is

$$\cos(x) = 1 - \frac{x^2}{2!} + \frac{x^4}{4!} - \ldots$$

Evaluate the Maclaurin cosine series using a `Do While` loop. Limit your series to 30 terms and a tolerance of $0.5e^{-6}$. Apply your series to $x = 28°$.

4.6 Use a `For` loop to compute the sum of the numbers from 1 to 50, their product, and the sum of the odd numbers from 1 to 50.

4.7 Repeat Exercise 4.5 but limit your series to five terms. Use a `For` loop instead of a `Do While` loop.

4.8 Evaluate the tan series from Exercise 3.8 in Chapter 3. Enter x and x_0 in degrees, the maximum number of terms, and the tolerance. Then convert x and x_0 to radians within your program. Use a `Do While` loop to evaluate your series. Continue the loop until a term less than the tolerance (in magnitude) is computed or the maximum number of terms has been reached. Output your answer, the exact value, the number of terms computed, the last term, and the true error. Let $x = 70°$ and $x_0 = 45°$. Limit your series to 30 terms and a tolerance of 0.5×10^{-6}.

4.9 A simple computer game is to try to guess a number picked at random by a computer. Assume the numbers must be integers between 1 and 1000, inclusive. Write a VBA program to play this game using `If` statements. After each guess, the computer will indicate whether the number is correct or whether the guess is high or low. It will keep track of the number of guesses. The computer picks

its number using the random number generator, which generates a number between 0 and 1. This number is then multiplied by 1000 and its integer value selected to produce the target number. Thus, `g = Abs(Int(1000 * Rnd(seed)))`
where `RND` is the VBA random number generator.

The computer can also play the game based on a number input by the
player. Its method of attack is to find the midpoint of the eligible range. Thus,
at the start, its first guess will be 500. If this is low, its next guess will be 750;
if high, 250, and so on.

4.10 The following shows some of the equations of projectile motion (without drag).
Vx and Vy are the x and y components of the velocity, Θ is the angle of the trajectory, subscript 0 denotes initial conditions, and t is the time.

$$Vx = V_0 * \cos(\Theta_0), \qquad Vy = V_0 \sin(\Theta_0) - g * t, \qquad \tan(\Theta) = \frac{V_y}{V_x}$$

$$x = x_0 + V_0 * \cos(\Theta_0) * t \qquad y = y_0 + V_0 \sin(\Theta_0) * t - .5 * g * t^2$$

Write a program to input the initial values of V and Θ (in degrees). Within
your program, convert Θ to radians. Then use a `For` loop to compute 20 steps
of the trajectory where the time is advanced 1 second for each step. Begin with
$t = 0$. Output your input and your results. Include a flowchart. Let $V_0 = 80 +$
t/s, $\theta_0 = 61°$, $x_0 = 0$, and $y_0 = 0$

4.11 The Maclaurin series expansion for the inverse tangent is

$$\tan^{-1}(x) = x - \frac{x^3}{3} + \frac{x^5}{5} - \frac{x^7}{7} \cdots$$

Write a VBA program to enter x. Use a `Do While` loop to evaluate your
series. Continue the loop until a term less than tolerance (in magnitude) is computed or the maximum number of terms has been reached. Evaluate your series for $x = 2$. Limit your series to 30 terms and a tolerance of 0.5×10^{-6}.
Output your input, your answer, the exact value, the number of terms computed, the last term, and the true error.

4.12 Bernoulli numbers are generated by the following formula:

$$B_{2n-1} = \frac{2[(2n)!]}{(2^{2n} - 1) * \pi^{2n}} * \left[1 + \frac{1}{3^{2n}} + \frac{1}{5^{2n}} + \frac{1}{7^{2n}} + \cdots \right].$$

Since $2n - 1$ is always odd, the Bernoulli numbers are only odd: that is,
B_1, B_3, B_5, and so on.

Write a VBA program to evaluate the first seven Bernoulli numbers. Use
a `For` loop in your main program to define which Bernoulli number you are generating. Within this loop, use a `Do While` loop to evaluate the series for the
value of n defined by your `For` loop. Continue evaluating a particular Bernoulli
number until a series term is less than or you have evaluated 30 terms. Calcu-

late the factorial using a loop within your `Do While` loop.

Input the number of Bernoulli numbers to evaluate, the tolerance, and the maximum number of terms. Output should be to a file and include the input values and the Bernoulli number calculated (e.g., 1, 3), its value, the number of terms used, and the value of the last term. Begin by making a flowchart.

4.13 Bessel functions may be generated from the following series where Γ is the Gamma function.

$$J_n(x) = \sum_{k=0}^{\infty} \frac{(-1)^k x^{n+2k}}{2^{n+2k} \, k! \, \Gamma(n + k + 1)}.$$

Note: $\Gamma(m) = (m - 1)!$ if m is an integer and $\Gamma(1) = 1$. Then $\Gamma(2) = (2 - 1)! = 1, \Gamma(3) = 2!$, and so on. Also, $0! = 1$ and $(-1)^0 = 1$.

Write a VBA program to input x, n, the maximum number of terms to be considered, and the tolerance. Then evaluate $J_n(x)$. Continue your series until a term becomes less than the tolerance or the maximum number of terms has been reached.

Begin by making a flowchart. Use a `Do While` loop to control the overall series and use `For` loops within it to calculate $k!$ and Γ.

Output should include the input values and the calculated values of $J_n(x)$, the number of terms used, and the value of the last term. Use your program to calculate the first three J_n's for $x = 1.5$; that is, J_0, J_1, and J_2. Control the values of n in an outer `For` loop. Continue evaluating the series for a given n until a term is less than 0.5×10^{-6} or you have evaluated 30 terms.

Compare your results to the published values. Cite the source of the published values.

5 | Numerical Integration

■ 5.1 The Basic Idea: Area

Numerical integration exploits a fundamental property of the mathematical process of integration: that is, that an integral represents the area under a curve. Schemes of numerical integration are simply ways of approximating this area. One simple but tedious and inaccurate way to do this would be to plot the function to be integrated on grid paper, establish a scale for the grid, sum the squares, and—using the scale—convert it to the desired units. We can do better than this.

The techniques to be discussed in this chapter apply to definite integration only. Thus, limits must be specified and the result produced is numerical. The integral must be continuous in the closed range of the limits. Closed-form integration, such as that performed by packages like Maple, will not be considered. The results of numerical integration are only approximations to the area, but formulas exist that provide good estimates for the errors in the integral approximation.

Consider the curve shown in Figure 5.1.

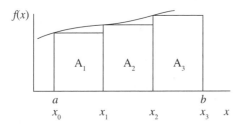

I FIGURE 5.1 Rectangular Areas under a Curve.

The area under the curve from $x = a$ to $x = b$ is defined by the integral

$$\int_a^b f(x)dx.$$

If we place rectangles A_1, A_2, A_3, and so on under the curve, we can evaluate each of these areas and then add them to determine the area under the curve. Of course, the rectangles do not fit too well, so the area calculated will not be a good approximation. We can easily make the approximation better by putting in more, but smaller, areas between a and b.

We must decide how many intervals or areas, n, we want to insert between a and b. Then the width of each area is defined by $\Delta x = \dfrac{(b - a)}{n}$. There is a specific relationship between n and Δx. Instead, we could have specified Δx and then $n = \dfrac{(b - a)}{\Delta x}$. If the spacing is varied between a and b, the integral must be broken into pieces, with a constant Δx associated with each piece.

It is standard practice to number the points from 0 to n and the intervals from 1 to n; thus, there is one more point than there are intervals. If the areas used are rectangles, each area will be given by $f(x_i) * \Delta x$. We could have defined the height of the first rectangle by either x_0 or x_1. The choice is arbitrary but leads to slightly different answers. Usually, we start with x_0.

■ 5.2 The Trapezoid Rule

Our results can be considerably improved if we add a triangle to the top of each rectangle approximating the curve. The slanted top does a better job fitting the curve than does the horizontal top of the rectangle, as shown in Figure 5.2. Then area A_i becomes the area of the rectangle, A_{ABCD}, plus the area of the triangle A_{DCE}.

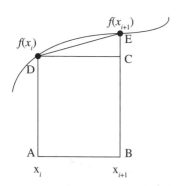

I FIGURE 5.2 Trapezoid Area under a Curve.

For the rectangle and triangle, let

$$A_i = f(x_i) * (x_{i+1} - x_i) + \frac{1}{2} * (x_{i+1} - x_i) * (f(x_{i+1}) - f(x_i))$$

$$= f(x_i) * \Delta x + \frac{1}{2} * \Delta x * (f(x_{i+1}) - f(x_i))$$

$$= \frac{f(x_i) + f(x_{i+1})}{2} * \Delta x.$$

This is the area of a trapezoid and the resulting formula is called the *trapezoid rule*. For the next area,

$$A_{i+1} = \frac{f(x_{i+1}) + f(x_{i+2})}{2} * \Delta x.$$

Thus, x_{i+1} appears twice, once in each of these two areas, which is true of all interior points. However, the endpoints appear only once. (The interior points are like the interior walls in a house that are common to two rooms.) Thus, the formula for the trapezoid rule is as follows:

$$\int_a^b f(x)dx = \frac{\Delta x}{2} * [f(a) + 2 * f(a + \Delta x) + 2 * f(a + 2 * \Delta x)$$

$$+ \ldots + 2 * f(b - \Delta x) + f(b)]. \tag{5.1.1}$$

Let us now apply the trapezoid rule to the integral

$$\int_1^5 x^2 \ln(x)dx.$$

Let $n = 4$ so $\Delta x = \frac{(b - a)}{n} = \frac{(5 - 1)}{4} = 1.$

It is a good idea to choose an easily computed interval such as $\frac{1}{2}$ or $\frac{1}{4}$ when making these calculations by hand. We could set up a table to facilitate computation.

x	f(x)	factor	product
1	$1^2 * \ln(1) = 0$	1	0
2	$2^2 * \ln(2) = 2.77259$	2	5.54518
3	$3^2 * \ln(3) = 9.88751$	2	19.77502
4	$4^2 * \ln(4) = 22.18070$	2	44.36142
5	$5^2 * \ln(5) = 40.23595$	1	40.23595
		Sum	109.91757

Then Integral I $= \frac{\Delta x}{2} *$ Sum $= \frac{1}{2} * 109.91757 = 54.95879.$

This integral, though not obvious, can be evaluated in closed form using integration by parts. Let $u = \ln(x)$ and $dv = x^2\, dx$. Then,

$$du = \frac{1}{x} dx \text{ and } v = \frac{x^3}{3}$$

$$\int_a^b x^2 \ln(x) dx = \int_a^b u\, dv = uv\, |_a^b - \int_a^b v\, du$$

$$\int_1^5 x^2 \ln(x) dx = \ln(x) * \frac{x^3}{3}\bigg|_1^5 - \frac{x^3}{9}\bigg|_1^5 = 53.28214.$$

Thus, the true error is $53.28214 - 54.95879 = -1.67665$.

A negative error means the estimated value is greater than the true value. Obviously, if we do not know the correct value of the integral (which we certainly shall not if we really must resort to numerical integration), we cannot compute the true error. Fortunately, there are straightforward formulas for estimating the error. These formulas will be presented now without proof. See McCracken and Dorn (1961) for a justification of the trapezoid rule error formula.

5.2.1 Trapezoid Rule Errors

E_{TRUN}, the truncation error is given by

$$E_{TRUN} = -\frac{\Delta x^2 * (b - a) * f^{II}(x)}{12}, \tag{5.2.1}$$

(handwritten margin note: Choose max f''(x) or avg. f''(x))

where $f^{II}(x)$ is the second derivative of the integrand over the interval $a <= x <= b$. It may seem strange that the error in the integral of $f(x)$ depends on its second derivative. This should immediately raise the question as to what value of x should be used. There are two reasonable answers.

5.2.1.1 The Maximum Error

The maximum possible error occurs if we use the largest value of $f^{II}(x)$ in the interval. This is analogous to bounding the error in a series (see Chapter 3, Section 3.3) and requires that we find the maximum value, in magnitude, of $f^{II}(x)$ in the range of x from a to b.

If $f(x) = x^2 * \ln(x)$,

$$f^{I}(x) = 2x * \ln(x) + \frac{x^2}{x} = 2x * \ln(x) + x,$$

and

$$f^{II}(x) = 2 * \ln(x) + 2\frac{x}{x} + 1 = 3 + 2 * \ln(x).$$

Since $\ln(x)$ increases monotonically as x increases, it is obvious that the maximum value occurs at $x = b = 5$. There, $f^{II}(5) = 3 + 2 * \ln(5) = 6.21888$.

Then, using Equation (5.2.1),

$$E_{TRUN,max} = \frac{-1^2 * (5 - 1) * 6.21888}{12} = -2.07296.$$

This is bigger (in magnitude) than the true error, and so we have successfully bounded the error. Although we usually do not care if an error is plus or minus, in integration the sign of the error can be important. For example, if the integral represents a volume and we are buying concrete to fill that volume, knowing whether the estimate is high or low can be economically important. We do not want to buy an insufficient amount and have to go back later to buy more, probably at a higher per unit cost!

5.2.1.2 The Average Error

Estimating the maximum error can be very conservative. Alternatively, we can compute the average error over the interval $a <= x <= b$. We can do this by computing the average value of the second derivative over this interval. This is easy to do. The average value of the Nth derivative of a continuous function over an interval is

"N" order of derivative → NOT "n" number of intervals

$$f^N(x) = \frac{(f^{N-1}(b) - f^{N-1}(a))}{b - a},$$

where f^{N-1} is the previous (next lower) derivative. Thus, if $N = 2$ as in the trapezoid rule error formula, $N - 1 = 1$, and we only need the first derivative of the integrand to estimate the error. Furthermore, we do not have to worry about where to evaluate the derivative. We use only the endpoints. Finally, the $(b - a)$ in the numerator of Equation (5.2.1) cancels the $(b - a)$ in the denominator of the $f^N(x)$ equation. This leads to the average truncation error formula for the trapezoid rule:

$$E_{TRUN,avg} = -\frac{\Delta x^2 * (f^{N-1}(b) - f^{N-1}(a))}{12} \tag{5.2.3}$$

By applying this formula to the present problem, the average truncation error can be determined as follows:

$$f^2(x)_{,avg} = \frac{(f^1(b) - f^1(a))}{b - a}$$

where

$$f^1(x) = 2x * \ln(x) + x$$
$$f^1(a) = f^1(1) = 2 * \ln(1) + 2 = 1$$

and

$$f^1(b) = f^1(5) = 2 * 5 * \ln(5) + 5 = 21.09438.$$

So,

$$E_{TRUN,avg} = -\Delta x^2 * \frac{f^1(5) - f^1(1)}{12}$$
$$= -1^2 * \frac{(21.09438 - 1)}{12} = -1.67453.$$

This is very close to the true error calculated for this problem, though it is slightly nonconservative (i.e., underestimates the magnitude of the error). That is the chance we take in evaluating the average error. On the other hand, it is easier to perform for the reasons cited above.

Which error estimate form should we use? The average error is easier to compute and generally close enough. In life we often go with averages: the average time to drive a certain distance, the average high temperature in July, the average low temperature in January, and so on. We know these average values can be exceeded. But we can also use an error estimate as a guide to how many intervals we ought to use in an integral evaluation. The result from equation (5.2.1) depends on Δx, the interval size, raised to the second power. If we reduce Δx by, say, halving it, the other terms in the formula are unaffected but Δx^2 decreases by a factor of 4. Hence, halving the interval reduces the error in the trapezoid rule by a factor of 4. If we repeated our calculation with $\Delta x = 0.5$ ($n = 8$), we would get a result of 53.70090 with a true error of 0.41877. This error is only 0.2498 times the $n = 4$ result.

5.2.1.3 Development of the Average Derivative Formula

That the average Nth derivative over an interval $[a, b]$ is the difference between the $N - 1$ derivatives at b and a divided by the difference $b - a$ can be seen from both an analytic and a geometric perspective.

Suppose we were to foolishly try to add up the infinite number of Nth derivatives between a and b. We might try to average the infinite sum

$$f^N_{avg} = \frac{\lim_{n \to \infty} \sum_{i}^{n} f_i^N}{n}.$$

But the infinite sum divided by the number of terms becomes the integral of the Nth derivative divided by the interval

$$\frac{\int_a^b f^N dx}{b-a} = \frac{\int_a^b \frac{df^{N-1}}{dx} dx}{b-a} = \frac{f^{N-1}(b) - f^{N-1}(a)}{b-a}.$$

Or, let us consider Figure 5.3 where a function f^{N-1} is plotted against x, including the interval a to b. If we connect the values of f^{N-1} at the endpoints divided by the distance between a and b and divide them, we get

$$\frac{f^{N-1}(b) - f^{N-1}(a)}{b-a},$$

which is $\dfrac{\Delta f^{N-1}}{\Delta x}$, the average slope of f^N.

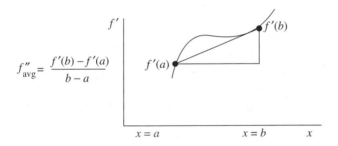

I FIGURE 5.3 Geometry of the Average Second Derivative.

■ 5.3 Simpson's 1/3 Rule, Something for Nothing

One way to improve the accuracy of the trapezoid rule results would be to increase the number of intervals. We have seen that doubling n reduces the error by a factor of 4. But another, more effective way is to use a method employing a parabola instead of a straight line to connect adjacent points defined by the intersection of the curve with the cells underneath. See Figure 5.4.

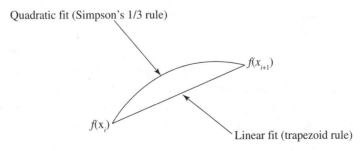

I FIGURE 5.4 Quadratic and Linear Connection of Two Points on a Curve.

Without deriving it, the formula for Simpson's 1/3 rule is

$$\int_a^b f(x)dx = \frac{\Delta x}{3} * [f(a) + 4 * f(a + \Delta x) + 2 * f(a + 2\Delta x)$$
$$+ 4 * f(a + 3\Delta x) + 2 * f(a + 4\Delta x) + \ldots$$
$$+ 2 * f(b - 2\Delta x) + 4 * f(b - \Delta x) + f(b)].$$

(5.3.1)

Note that, except for $f(a)$ and $f(b)$, the multipliers for the terms are alternating 4's and 2's. Also n, the number of intervals, *must* be an even number. If n is odd, an alternate formulation is available.

Let us apply Simpson's 1/3 rule to the previous integral with $n = 4$.

x	f(x)	factor	product
1	$1^2 * \ln(1) = 0$	1	0
2	$2^2 * \ln(2) = 2.77259$	4	11.09036
3	$3^2 * \ln(3) = 9.88751$	2	19.77502
4	$4^2 * \ln(4) = 22.18070$	4	88.72284
5	$5^2 * \ln(5) = 40.23595$	1	40.23595
		Sum	159.82417

Then $\int_a^b f(x)dx = \frac{\Delta x}{3} * \text{Sum} = \frac{1}{3} * 159.82417 = 53.27472.$

Recalling that the correct answer, as obtained in Section 5.2, is 53.28214, we have a true error $= 53.28214 - 53.27472 = 0.00742$, which is considerably better than with the trapezoid rule using the same n, which required about the same amount of work. But we must pay attention to the alternating factors in the expression and remember that n must be even!

The truncation error formula for Simpson's 1/3 rule is given by the formula

$$E_{\text{TRUN}} = \frac{\Delta x^4 * (b - a) * f^{IV}(x)}{180}.$$

(5.3.2)

Notice several things. First, it is of the same form as the truncation error formula for the trapezoid rule but the interval size is to the fourth power. This means that halving the interval will reduce the error by a factor of 16. Second, the error is proportional to the fourth derivative of the integrand, $f(x)$. This means that Simpson's 1/3 rule is exact for a cubic integrand (whose fourth derivative would be 0). Since the method is derived by passing a parabola through adjacent points, this is surprising and better than we could hope for. We get cubic accuracy with a parabolic fit, that is, something for nothing, which does not happen very often. (The reason is that some terms in the derivation are equal but with the opposite sign so they cancel each other.)

Let us estimate the error using this formula. Recall that $f^{II}(x) = 2 * \ln(x) + 3$.

Then, $f^{III}(x) = \dfrac{2}{x} = 2x^{-1}$ and $f^{IV}(x) = -2x^{-2}$.

Again, we may calculate either the average or the maximum error, depending on how we evaluate the fourth derivative. We shall consider the average error first, which requires that we evaluate the average fourth derivative. As with the trapezoid rule, average error evaluation of the average fourth derivative is given by

$$f^{IV}_{avg} = \frac{f^{III}(b) - f^{III}(a)}{b - a}$$

$$= \frac{\frac{2}{5} - \frac{2}{1}}{5 - 1} = -0.4.$$

(handwritten annotation:)
$$f(x) = x\,e^{-x}$$
$$f'(x) = e^{-x} - xe^{-x} = e^{-x}(1-x)$$
$$f''(x) = -e^{-x}(1-x) - e^{-x} = e^{-x}(x-2)$$
$$f'''(x) = -e^{x}(x-2) + e^{-x} = e^{x}(3-x)$$

Then, $E_{TRUN,avg} = -\dfrac{1^4 * (5 - 1) * (-.4)}{180} = 0.00889.$

This is close to and slightly exceeds the true error value of 0.00742.

If we compute the maximum error, we must look at the behavior of the fourth derivative. As x increases, $\dfrac{2}{x^2}$ decreases, so we will evaluate the maximum fourth derivative at $x = a = 1$ (again, we are concerned with the magnitude in locating the maximum). Then $f^{IV}_{max} = -\dfrac{2}{1^2} = -2$, and the error is estimated to be

$$E_{TRUN,max} = -\frac{1^4 * (5 - 1) * (-2)}{180} = 0.0444, \quad = \frac{-\Delta x^4}{180}\left[f'''(b) - f'''(a) \right]$$

considerably larger than the true error. That is reasonable, since we were using a conservative approach.

■ 5.4 Simpson's 3/8 Rule

Sometimes it is necessary to evaluate an integral with an odd number of intervals. This is most likely to happen if we are integrating measured data rather than a function (see Section 5.7), but we might also have an odd number if we want to use an easy-to-work-with interval. Suppose we wanted to apply Simpson's rule to the integral we have been considering but with the upper limit of 8 instead of 5 and still wanted to use $\Delta x = 1$. Then,

$$n = \frac{(b - a)}{\Delta x} = \frac{(8 - 1)}{1} = 7.$$

We could break this integral into parts with an even number of intervals plus three. Simpson's 3/8 rule could then be applied to the three-interval part. Note Simpson's 3/8 rule works *only* for $n = 3$, not for any other odd number. But any odd number greater than 1 can be broken into an even number plus 3. So this is not a serious limitation. Therefore, with $n = 8$, the integral can be broken apart as follows:

$$\int_1^8 x^2 \ln(x)dx = \int_1^5 x^2 \ln(x)dx + \int_5^8 x^2 \ln(x)dx.$$

Notice that 5 is the upper limit of the first part and the lower limit of the second. It is used twice. We also could have broken it up into 1 to 4 ($n = 3$) and 4 to 8 ($n = 4$). The present breakup makes use of the integral from 1 to 5 that we have already evaluated.

Simpson's 3/8 rule is as follows:

$$\int_a^b f(x)dx = \frac{(b - a) * [f(a) + 3 * f(a + \Delta x) + 3 * f(a + 2\Delta x) + f(b)]}{8}. \quad (5.4.1)$$

Applied to the integral from 5 to 8, Simpson's 3/8 rule yields

$$\int_5^8 x^2 \ln(x)dx = \frac{(8 - 5) * [f(5) + 3 * f(6) + 3 * f(7) + f(8)]}{8}.$$

So,

$$\int_5^8 x^2 \ln(x)dx = \frac{(8 - 5) * [40.23595 + 3 * 64.50334 + 3 * 95.34960 + 133.08426]}{8}$$

$$= 244.82963.$$

Note that a and b have different values from the first interval that went from 1 to 5. Then,

$$\int_1^8 x^2 \ln(x)dx = 53.27472 + 244.82963 = 298.10435.$$

Using integration by parts as discussed in Section 5.1, the correct answer is 298.11358. Then the true error of our numerical result is 298.11358 − 298.10435 = 0.00923.

The truncation error formula for Simpson's 3/8 rule is

$$E_{TRUN} = -\frac{(b-a)^5 * f^{IV}(x)}{6480}. \tag{5.4.2}$$

Finding the average error first, we recall that $f^{III}(x) = \dfrac{2}{x}$.

Then,

$$f^{IV}{}_{avg} = \frac{(f^{III}(b) - f^{III}(a))}{(b-a)} = \frac{\dfrac{2}{8} - \dfrac{2}{5}}{8-5} = -0.05,$$

and

$$E_{TRUN,avg} = -\frac{-(8-5)^5 * (-.05)}{6480} = 0.00188.$$

Combined with the average error for the interval from 1 to 5, we obtain

$$E_{TRUN,avg} = 0.00889 + 0.00188 = 0.01077.$$

This is slightly larger than the true error, so it is a conservative result.

To evaluate the maximum error of the integration from 5 to 8, we again need the fourth derivative, $f^{IV}(x) = -\dfrac{2}{x^2}$. Since this decreases in magnitude as x increases, the maximum magnitude occurs at $x = 5$ where $\dfrac{-2}{5^2}$ is −0.08. Then,

$$E_{TRUN,max} = -\frac{(8-5)^5 * (-.08)}{6480} = 0.00300.$$

Combined with the maximum error for the interval from 1 to 5, we obtain

$$E_{TRUN,max} = 0.04444 + 0.00300 = 0.04744.$$

This bounds (exceeds) the true error by a considerable amount.

■ 5.5 Richardson Extrapolation: A Clever Idea

In principle, we could reduce the truncation error by repeatedly increasing the number of intervals, thereby reducing the interval size. On the other hand, this increases the number of calculations made and therefore the round-off error. Eventually the round-off error may increase faster than the truncation error decreases (see Figure 22.2, Chapra and Canale, 2005). Therefore, it would be nice to find a way to increase the accuracy without using many more intervals. Richardson found a way.

Recall that the truncation error in the trapezoid rule is given by

$$E_{TRUN} = -\frac{\Delta x^2 * (b - a) * f^{II}(x)}{12}.$$

For a given integral, as we change n all of the terms are constant except Δx. Thus, we may write E_{TRUN} as $C * \Delta x^2$. Suppose we do an evaluation of a given integral using a particular $\Delta x = \Delta x_1$ to obtain I_1. The result including the error term would be

$$I = I_1 + C * \Delta x_1^2.$$

If we repeat the calculation with another interval, Δx_2, we would obtain

$$I = I_2 + C * \Delta x_2^2.$$

Since the value of the actual integral would not change, these two expressions could be equated to one another and the resulting equation solved for C. If we further specify that $\Delta x_1 = 2 \Delta x_2$ —that is, the number of intervals is doubled between the two calculations—a little algebra yields the result

$$I = \frac{[4 * I_2 - I_1]}{3}, \tag{5.5.1}$$

where I_2 is computed with the smaller Δx (greater n) and is hence more accurate. This is Richardson's extrapolation formula. It is based on the trapezoid rule.

Let us apply this to the previous integral. With $\Delta x = 1$, we found that $I_1 = 54.95879$. If the calculation is repeated with $\Delta x_2 = 0.5$, we would obtain $I_2 = 53.70091$.

Combining these two results using Richardson's formula, we obtain

$$I = \frac{[4 * 53.70091 - 54.95879]}{3} = 53.28162.$$

Recalling that the correct answer is 53.28214, we now have a true error of only 0.00052. This is a significant improvement requiring little extra work and only eight intervals. Richardson succeeded in improving the error without incurring the penalty of a round-off error increase when the number of intervals increases considerably.

That Richardson extrapolation is effective, can be demonstrated dramatically by combining crude results of trapezoid rule integration applied to our integral. If $n = 1, \Delta x = 4$, and we would obtain $I = 80.47190$. Since the correct value is 53.28214, we have a true error of -27.18976, a lousy result. If $n = 2$, we would obtain 60.01097, which yields a true error of -6.72883, a merely poor result. (Try these for yourself. It should not be surprising that the results are not good since Δx is large.)

But if we combine these two results using Richardson extrapolation ($n = 2$ is the more accurate because its Δx is smaller), we obtain

$$I = \frac{[4 * 60.01097 - 80.47190]}{3} = 53.19066,$$

which has a true error of only 0.09148, and a true relative error of 0.17%. This is surprisingly good.

■ 5.6 Romberg Integration

If combining results using Richardson extrapolation is good, why not continue the process? If we use the trapezoid rule for our standard integration with $n = 1, n = 2, n = 4$, and $n = 8$, we can combine these results, two at a time, using Richardson's extrapolation, to get three, more improved results. These new results, $k = 2$ (k is the column), are:

n	Δx	$k = 1$ trapezoid	$k = 2$ Richardson
1	4	80.47190	
			53.19066
2	2	60.01097	
			53.27473
4	1	54.95879	
			53.28160
8	0.5	53.70091	

Now if we apply Romberg integration, we can combine the results in the Richardson column ($k = 2$) to get another column containing two results. The rule for this column ($k = 3$) is

$$I_{1,3} = \frac{16 * I_{2,2} - I_{1,2}}{15} = \frac{16 * 53.27473 - 53.19066}{15} = 53.28034.$$

Likewise, the next entry in the third calculated column, $I_{2,3} = 53.28206$, which comes from combining $I_{3,2}$ and $I_{2,2}$. Thus, column $k = 3$ would have the following results:

$k = 3$
53.28034
53.28206.

Finally, we may combine these two results to obtain

$$I_{1,4} = \frac{64 * I_{2,3} - I_{1,3}}{63} = \frac{64 * 53.28206 - 53.28034}{63} = 53.28209.$$

The general rule of Romberg integration underlying these calculations can be applied as follows:

$$I_{r,k} = \frac{4^{k-1} * I_{r+1,k-1} - I_{r,k-1}}{4^{k-1} - 1} \tag{5.6.1}$$

where r is the row and k is the column.

Richardson extrapolation is thus the first column of a process that may be extended indefinitely to the right. The number of columns is equal to the number of rows in the first column. The truncation error for the trapezoid rule is proportional to Δx^2. This is in column $k = 1$. For Richardson extrapolation (the second column), the error is proportional to Δx^4, the same order of error as with Simpson's 1/3 rule. For $k = 2$, Δx^4 may also be written as Δx^{2k}. This relation holds for all the columns in Romberg integration. Thus, the only entry in our fourth column ($k = 4$) has a truncation error proportional to Δx^8.

A flowchart for a Romberg integration program is found in Chapter 10.

5.7 Integration of Data

Up to now, we have integrated a function. But sometimes it is necessary to integrate data. For example, if we integrate the acceleration, we can obtain the change in velocity. If we integrate a second time, we can compute the change in position. In fact, inertial navigation systems, which were used for airplane navigation across oceans before global positioning systems, electronically integrated the accelerations measured by inertially resistant spinning gyroscopes on the airplane.

Since we are integrating existing data instead of functions, we do not have to worry about evaluating the integrand at the intervals. However, we must use the data where given, and so we lose control of the spacing and number of intervals. As long as we follow the rules on placement (for example, n must be even when using Simpson's 1/3 rule), we may use any method we want.

Let us integrate the following data for acceleration as a function of time to obtain the change in velocity. First, note that

$$\frac{dV}{dt} = a$$

so

$$a\,dt = dV.$$

Then,

$$\int_{t0}^{t1} a\,dt = \int_{t0}^{t1} dV = V(t_1) - V(t_0).$$

Notice that we are computing only the change in V. If we want the value of $V(t_1)$, we must know $V(t_0)$, the initial value of V.

Suppose we are given the following data:

t	a
0	1.05
1	1.65
2	2.10
3	2.40
4	2.60
6	2.95
8	3.08
10	3.40
12	3.51
14	3.75

Two things can be noticed immediately. From $t = 0$ to $t = 4$, the time interval, Δt, is 1.0, and there are 4 intervals. From $t = 4$ to $t = 14$, the time interval is 2 and there are 5 intervals. Thus, we must break up our integration to conform to the time interval change, and if we want to use Simpson's rules, we must break up $n = 5$ into $n = 2$ plus $n = 3$ (or the other way around). Accordingly,

$$\int_{0}^{14} adt = \int_{0}^{4} adt + \int_{4}^{8} adt + \int_{8}^{14} adt$$

1/3 rule + 1/3 rule + 3/8 rule

$n = 4$ $n = 2$ $n = 3$.

With Simpson's 1/3 rule and $n = 4$, we obtain

$$\int_{0}^{4} adt = \frac{\Delta t}{3} * [a(0) + 4 * a(1) + 2 * a(2) + 4 * a(3) + a(4)],$$

then,

$$V(4) - V(0) = \frac{1}{3} * [1.05 + 4 * 1.65 + 2 * 2.10 + 4 * 2.40 + 2.60] = 8.02.$$

With Simpson's 1/3 rule and $n = 2$, we obtain

$$\int_{4}^{8} adt = \frac{\Delta t}{3} * [a(4) + 4 * a(6) + a(8)],$$

then,

$$V(8) - V(4) = \frac{2}{3} * [2.60 + 4 * 2.95 + 3.08] = 11.65.$$

With Simpson's 3/8 rule we obtain the following:

$$\int_{8}^{14} adt = \frac{(b - a)}{8} * [a(8) + 3 * a(10) + 3 * a(12) + a(14)],$$

then,

$$V(14) - V(8) = \frac{(14 - 8)}{8} * [3.08 + 3 * 3.40 + 3 * 3.51 + 3.75] = 20.67.$$

The total change in V is $V(14) - V(0) = 8.02 + 11.65 + 20.67 = 40.34$.

We have had to change spacing and the rule used to accommodate the given data, but the actual process is easier than when integrating a function because we do not have to perform any function evaluations. Note that we only computed the change in velocity from $t = 0$ to $t = 14$. To obtain the absolute velocity at t = 14, we would have to know the value of $V(0)$.

■ 5.8 The Extended Midpoint Rule

Sometimes an integral is encountered in which a division by zero may occur, but the integral is nonetheless finite. Without going into the details, such an integral exists if the integrand is everywhere equal to, or smaller than, an integrand for an integral that does exist over the domain of the independent variable. If the techniques we have used up to now will not work for such an integral because of the singularity, we may be able to evaluate the integral by avoiding the function at the trouble spot.

Consider the following integral that was encountered in a heat transfer problem involving flow over a flat plate with an insulated tip. Calculating the convective heat transfer coefficient required evaluating the integral. Note that the integrand "blows up" at the lower limit where $x = 2$:

$$\int_2^3 \frac{0.5x^{-0.5}dx}{(1 - (2/x)^{0.75})^{(1/3)}}.$$

The extended midpoint rule uses rectangles to approximate the function of the integrand. It thus has a relatively large truncation error. But instead of evaluating the integrand at the beginning or the end of the interval, we evaluate it at the midpoints of the intervals: that is, at $x = a + \dfrac{\Delta x}{2}$, $x = a + 3\dfrac{\Delta x}{2}$, and continuing out to $x = b - \dfrac{\Delta x}{2}$. Since the width of each rectangular interval is Δx and the height is $f(x_{i+1/2})$, the integral is approximated as

$$\int_a^b f(x)dx = \Delta x \sum_{i=0}^{n-1} f(x_{i+\frac{1}{2}}). \tag{5.8.1}$$

Thus, we have a rectangle rule, with the height of each rectangle defined at the middle of the interval.

Suppose we select $n = 4$ and evaluate the above integral with the extended midpoint rule. Then $a = 2$, $b = 3$, and $\Delta x = 0.25$.

x	f(x)
2.125	0.96828
2.375	0.65609
2.625	0.54210
2.875	0.47565
Sum	2.64212

Then the integral $= \Delta x * \text{Sum} = 0.25 * 2.64212 = 0.6605$.

This result, obtained with n only equal to 4, is crude. The correct answer is 0.7096, so more intervals would be needed to improve the result; nonetheless, the principle is illustrated.

■ 5.9 Chapter 5 Exercises

5.1 Consider the following integral:

$$\int_0^{0.5} (1 - x^2)^{0.5} dx.$$

a. Evaluate this integral with $n = 1$, $n = 2$, $n = 4$, and $n = 8$ intervals, using the trapezoid rule.

b. Estimate the average and the maximum errors when $n = 4$, and $n = 8$.

5.2 a. Evaluate again the integral in Problem 5.1 with Simpson's 1/3 rule with $n = 4$.

b. Estimate the average error for your Simpson's 1/3 rule result.

5.3 Consider the integral in Exercise 5.1 but replace the upper limit with 0.875.

a. Evaluate the integral using Simpson's rule(s) with $\Delta x = 0.125$ throughout.

b. Estimate the average error of the complete integral.

5.4 Use your trapezoid rule results from Exercise 5.1 to apply Romberg integration as far as possible.

5.5 Consider the following integral:

$$\int_0^{\pi/2} e^x * \cos(x) dx, \text{ } x \text{ in radians.}$$

a. Evaluate the integral using the trapezoid rule with $n = 1, 2, 4,$ and 8 intervals.

b. Estimate the maximum error with $n = 4$.

c. Using the results from Exercise 5.5a, apply Romberg integration as far as you can.

5.6 a. Evaluate again the integral in Exercise 5.5 with Simpson's 1/3 rule with $n = 4$.

b. Estimate the maximum and average errors for your results.

5.7 a. Evaluate again the integral in Exercise 5.5 with $\Delta x = \dfrac{\pi}{8}$ but use $\dfrac{7\pi}{8}$ as the upper limit of the integral. Use Simpson's rule(s).

b. Estimate the average error.

5.8 a. Evaluate the following integral using the trapezoid rule with $n = 1, 2, 4$, and 8 intervals:

$$\int_0^1 2^x \, dx.$$

b. Estimate the maximum and average errors.

c. Apply Romberg integration to your results from Exercise 5.8a. as far as possible.

5.9 a. Repeat Exercise 5.8 with Simpson's 1/3 rule and $n = 4$.

b. Compute the maximum and average errors.

5.10 Evaluate the integral of Problem 5.1 using Simpson's 1/3 rule with $a = 0$, $b = 1.25$, and $n = 14$. Use a Microsoft Excel Spreadsheet.

5.11 Apply the extended midpoint rule to the following integral, encountered in a dynamics class. Use $n = 16$ with Microsoft Excel.

$$\int_0^{100} \frac{ds}{(12s + 0.2s^2)^{0.5}}.$$

5.12 The integral $\dfrac{2}{\sqrt{\pi}} \displaystyle\int_0^z e^{-(t^2)/2} dt$ is sometimes called the probability integral. If $z = 0.8$, evaluate this integral with the trapezoid rule and 4 intervals. Estimate the average and maximum errors.

5.13 a. Repeat Exercise 5.12 with Simpson's 1/3 rule and $n = 4$.

b. Compute the maximum and average errors.

5.14 Write a program to evaluate the following integral with the trapezoid rule with 16 intervals:

$$\int_0^1 \frac{0.002 * (4 * x^2 - x^3) dx}{(0.1 + \sin(0.063 * x))}.$$

5.15 Write a program to evaluate the integral in Exercise 5.14 using Simpson's 1/3 rule with the trapezoid rule with 16 intervals.

5.16 The following data were obtained for the variation of acceleration with time:

t, secs	a, ft/sec²
0.0	0.00
1.0	1.01
2.0	7.98
4.0	26.5
6.0	65.0
8.0	120.00

Given that $V(0) = 0$, calculate the object velocity at $t = 8.0$ secs. Use Simpson's rule(s).

5.17 Consider the following velocity data as a function of time:

t, secs	V, ft/sec
0	0
1	1.39
2	3.30
3	5.55
4	8.05
6	13.6
8	19.8
10	26.4
12	33.3

Determine the displacement between $t = 0$ and $t = 14$ by integrating these data. Use Simpson's 1/3 rule.

5.18 The following data were provided for the variation of pressure with volume.

V, in³	P, lb/in²	V, in³	P, lb/in²
13.0	112	5.0	270
11.0	131	3.0	424
9.0	157	2.0	500
7.0	197	1.0	525

Given that work $= \int P\, dV$, calculate the work performed in this process in ft-lb by integrating these data by hand. Use Simpson's rule(s).

6 Subprograms and Functions: Useful Specialists

■ 6.1 Why Subprograms and Functions?

Not infrequently in writing a program we have occasion to perform the same calculation more than once within that program. For example, we may need to solve different quadratic equations several times within a program. Each time we must do this, it would be possible to include the lines of code we developed in Chapter 4. But it would be more efficient if we included that coding only once somewhere in our program and go to that code location whenever we need to solve a quadratic equation. We would then return to the part of the program from which we made this temporary detour.

Subprograms and functions (they are very similar as we shall soon see) are special-purpose parts of a program. They are a bit like the special teams that enter football games when it is necessary to punt or receive a kick-off. They may be developed and debugged separately and then copied into another program. They are very handy.

■ 6.2 Subprogram Form

A subprogram is accessed (or called) where needed in the main program (or from another subprogram or function) by using a call statement of the form

```
Call Sub1(.......)
```

where `Sub1` is the arbitrary name of the subprogram, and the list within the parentheses (called the argument) is the list of variables needed in the subprogram or results returned from it to the calling program. Subprogram and function names follow the same rules as for variables; that is, they must begin with a letter but can then include letters, numbers, _ , and % after that. They cannot contain breaks or spaces, and must be no more than 255 characters long.

Each subprogram must be defined completely outside other subprograms or functions. Each must begin and end before another is defined. Typically these are defined, one after the other, below the main subprogram. Subprograms can call other subprograms or even call themselves. The latter is called *recursion*.

To solve a quadratic equation, we would need to send the subprogram the values of a, b, and c, and send back the values of the roots. We might also want to include an indicator to show whether the roots are real or complex. Thus, we might have the line in our program:

```
Call Quad(a1,b1,c1,r1,r2,ind)
```

Here a1, b1, and c1 are the equation coefficients, r1 and r2 are the roots (or real and imaginary parts if the roots are complex), and ind is a variable to indicate whether the solution is real or complex. The ind variable is needed to properly interpret and output the results. After the subprogram has been executed, control of the program returns to the line immediately after the Call line.

Somewhere lower in the program we might have a similar line:

```
Call Quad(a2,b2,c2,r3,r4,ind)
```

This time we are sending coefficients a2, b2, and c2 to the subprogram. That we have used different labels is not a problem.

The subprogram would begin with Sub and end with End Sub as

```
Sub Quad(a, b, c, x1, x2, flag)
    . . . .
    . . . .
End Sub
```

The program statements would be contained between these lines. The labels in the argument list of the subprogram may be different from those in the calling program. We can do this because the subprogram does not "read" the names of the variables. It simply takes them in the order in which they are specified in the calling program. Thus, a1 and later a2 in the calling program correspond to a in the subprogram, b1 and b2 to b, and so on. Hence, we must be careful that we preserve the order of variables. Also, we should always use variable names in the argument list rather than numbers, even if we have to create a name just for that purpose. Otherwise, we could literally change the value of a number in the subprogram. But we may use any names we want within the subprogram to identify variables, consistent with their names in the argument list. The argument list can contain both input from the calling program (e.g., a, b, and c) and results being sent back (x1, x2, and flag).

Flag is a flag or indicator, defined in this example to control output in the calling program instead of within the subprogram. Since the roots of a quadratic equation may be either real or complex, the form of the output is different in the two cases. This information would be needed in the calling program if results were output there. If d is the discriminant we might include logic like the following in the Quad subprogram:

```
if d >= 0 then
     flag = 1
. . . . . .
. . . . . .
Else
     flag = 2
. . . . .
. . . . .
endif
```

Output in the calling program would then assume the form of real roots if flag (or ind in the calling program) is 1, and the complex form if it is 2.

All variables needed in subprograms or functions must be either passed in the argument list or defined within the subprogram or function. Otherwise, their values will be unknown, and VBA will set them to zero.

6.2.1 Passing by Reference and Passing by Value

When we type array lists as we have just done, what we have actually done is passed the storage locations of the variables in the list to the subprogram. Thus,

```
Call Quad(a,b,c,r1,r2,ind)
```

passes the locations of a, b, and c from the calling program to the subprogram Quad. There, the variable values may be changed and the new values stored in the corresponding locations known to both the calling and called programs.

Occasionally, we may want to transmit only a copy of the value (but not the storage location) of the variable to the subprogram, leaving the original value intact. This is called *passing by value*. It is accomplished by enclosing the variable name(s) so passed in parentheses. Thus, if we typed

```
Call Quad((a),b,(c),r1,r2,ind)
```

only copies of the values of a and c would be passed to the subprogram, not their storage locations. This would protect their original values in the calling program.

6.2.2 Variable Typing in Subprograms

Variables listed in the argument list of the calling subprogram are typed in that subprogram. New variable names introduced within a subprogram are typed within that subprogram. See the example in Section 6.4

■ 6.3 Function Form

An alternate to using a subprogram is to use a function. These are like the library functions available in VBA (e.g., the trig and log functions) but are created by the programmer. Unlike subprograms, they return only a single result. For example, a function to solve a quadratic equation would return only one of the two roots. If the solution were complex, a single function could return only part of the solution. Thus, functions tend to be written and used for simpler calculations than subprograms. Like subprograms, each function must be defined completely outside other subprograms or functions. Each must begin and end outside the others.

Functions are defined according to the following form:

```
Function name(argument list)
    .....
    .....
    name =
End Function
```

The variable whose value is returned is the name of the function. Functions should not be typed. VBA treats them as variants. New variable names introduced within a function should be typed.

If we were to use a function to find two real roots of a quadratic equation, we could write

```
Function x1(a,b,c)
    .....
    .....
    x1 = (-b + sqr(b^2 - 4*a*c)/(2*a)
End Function
```

and

```
Function x2(a,b,c)
    .....
    .....
    x2 = (-b - sqr(b^2 - 4*a*c)/(2*a)
End Function
```

These functions could be called as

```
root1 = x1(a,b,c)
root2 = x2(a,b,c)
```

6.3.1 Static Subprograms and Functions

After a subprogram or function has been executed, variable values appearing only within them are erased. If we wish to retain them for the next time the subprogram or function is executed, we may define that program segment as static. For example, suppose we wish to keep track of how many times Sub Try is executed by incrementing a variable count each time the subprogram is called. Clearly we don't want this reset. We could retain the value of local variables such as count by prefixing the Sub Try definition with Static as:

```
Static Sub Try(.........,count)
      .....
      .....
      count = count +1
End Sub
```

The value of count would be retained from one call to the next.

■ 6.4 Example: The Trapezoid Rule

A subprogram to employ the trapezoid rule could use a function call to define the integrand. Assume that the limits of integration, a and b, have already been input, and the number of intervals defined. Let the integrand be $x^2 * \ln(x)$ as in the example in Chapter 5. Then we could have

```
Sub Main
Dim a as double, b as double, integ as double
Dim n as integer
.....
Call Trap(a, b, n, integ)
      .....
      .....
End Sub

Sub Trap(a, b, n, integ)
Dim dx as double, sum as double, x as double, i as integer
dx = (b - a)/n
sum = f(a) + f(b)
x = a
For i = 1 To n - 1
      x = x + dx
      sum = sum + 2*f(x)
Next i
integ = dx/2*sum
End Sub
```

```
Function f(x)
f = x^2*log(x)
End Function
```

If a different integrand were to be evaluated, it would only be necessary to redefine f in the function, and input the desired values of a, b, and n. The rest of the routine would be intact. This shows the power of using functions, for although Trap calls it repeatedly, f is only defined once.

■ 6.5 Unused Arguments

Not all of the variables named in a subprogram or function argument list need to be used. For example, suppose a function is used to define a differential equation for a spring-mass system with damping and a forcing function. The equation might be as follows:

$$\frac{dv}{dt} = Fo/m * \sin(wt) - c/m * v - k/m * x,$$

where v is the mass velocity, Fo the amplitude of the forcing function, m the mass, w the frequency of the forcing function, c the damping coefficient, and k the spring constant. A function for this differential equation could be the following:

```
Function f(v, Fo, m, w, c, k, x, t)
    f = Fo/m*sin(w*t)-c/m*v-k/m*x
End Function
```

In another problem, the differential equation might simply be

$$\frac{dv}{dt} = g.$$

We could use the same functional form listing all the arguments, but only using one (with g replacing Fo), as shown:

```
Function f(v, g, m, w, c, k, x, t)
    f = g
End Function
```

Although eight variables were passed to the function, only one was used. This is acceptable, but we must be careful not to use variables that are not passed or evaluated within the function. The ability to ignore variables in an argument list means we do not have to change its form each time we modify a particular function.

■ 6.6 Excel Functions in VBA

Many of the functions in the extensive Excel function library can be used in VBA programs. This may save considerable time and effort in solving problems that involve standard functions. The format is

```
Application.Worksheetfunction.name of function
```

The factorial function in Excel is `fact(n)`. If n has been defined within a VBA program, n! may be found by including the line

```
nfact = Application.Worksheetfunction.fact(n)
```

The list of Excel worksheet functions may be viewed from a pull-down menu that appears after `Worksheetfunction` has been entered in the VBA line. Double clicking on any of these will add it after the period following `Worksheetfunction`.

A list of Excel functions available in VBA is found in Appendix F.

■ 6.7 VBA Functions in Excel

It is possible to use VBA to create your own functions that can then be accessed by Excel. Suppose you wish to create a function to find the quality of a two-phase mixture of steam and water at a given pressure from the total enthalpy, h, the steam enthalpy, hg, and the liquid enthalpy, hf at that pressure. The formula is

$$x = \frac{(h-hf)}{(hg-hf)}.$$

Suppose h is in C1, hf is in C2, and hg is in C3. x could be calculated in C4 by creating a function *using the VBA editor* (not a macro) as:

```
Function x(h,hf,hg)
    x = (h-hf)/(hg-hf)
End Function
```

In C4 we could enter $=x(C1,C2,C3)$

Obviously this can also be done by simply typing the formula directly in C4. On the other hand if this formula were needed several times within the spreadsheet at different locations, defining the function in VBA once and simply calling it as needed would be simpler. Note we passed the values from the spreadsheet to the function in an argument list, as we would within a standard VBA program. More complicated functions could similarly be created. Variables should not be typed since their values are imported from the spreadsheet.

■ 6.8 Chapter 6 Exercises

6.1 Write a VBA program that finds the roots of a quadratic equation. Input the equation coefficients in the main program but put the solver in a subprogram. Output results from the main program. Apply your program to $3x^2 + 2x + 1 = 0$ and $3x^2 + 2x - 1 = 0$.

6.2 Consider three-dimensional vectors of the form $\mathbf{A} = \mathbf{A}_x\mathbf{i} + \mathbf{A}_y\mathbf{j} + \mathbf{A}_z\mathbf{k}$ and $\mathbf{B} = \mathbf{B}_x\mathbf{i} + \mathbf{B}_y\mathbf{j} + \mathbf{B}_z\mathbf{k}$. Write a program to compute:

a. the magnitude of $\mathbf{A}$,

b. a unit vector in the direction of $\mathbf{B}$,

c. the scalar (dot) product of $\mathbf{A}$ and $\mathbf{B}$, and

d. the vector (cross) product of $\mathbf{A}$ and $\mathbf{B}$.

Evaluate Part (a) with a function, and Parts (b), (c), and (d) in subprograms. Let

$\mathbf{A} = 3\mathbf{i} + 4\mathbf{j} + 5\mathbf{k}$ and $\mathbf{B} = -1\mathbf{i} + 10\mathbf{j} - 11\mathbf{k}$

6.3 Following the example in Section 6.4, write a VBA program to evaluate an integral using the trapezoid rule. Input the integral limits and the number of intervals into the main program. Then call a subprogram to apply the trapezoid rule. Evaluate the integrand in a function. Return the answer to the main program.

Apply your program to this integral:

$$\int_0^{\pi/2} e^x * \cos(x)dx, \ x \text{ in radians.}$$

6.4 Modify the program written for Exercise 6.3 to put calls to the trapezoid rule inside a Do While loop. Halve the interval size each time the loop executes. Remain in the Do While loop until the approximate relative error computed from successive calls, as a percentage, is less than 0.5×10^{-6}.

Apply your program to the integral in Exercise 6.3.

6.5 The law of sines is $\dfrac{a}{\sin(A)} = \dfrac{b}{\sin(B)}$. The law of cosines is $c^2 = a^2 + b^2 - 2 * b * a * \cos(C)$. Write a VBA program to input a, b, and B in the main subprogram. Put the law of sines in one function and the law of cosines in another. Use your program to compute the length of side c if $a = 85$, $A = 40°$, and $B = 30°$. Output from the main subprogram.

6.6 Work Exercise 6.5 using two additional subprograms instead of functions.

6.7 The sine of the sum of two angles is given by $\sin(A + B) = \sin(A) * \cos(B) + \sin(B) * \cos(A)$. The cosine of the sum of two angles is given by $\cos(A + B) = \cos(A) * \cos(B) - \sin(A) * \sin(B)$. Write a VBA program to input A and B in the main subprogram. Put the sine sum formula in one function and the cosine sum formula in another. Use your program to compute the sine and cosine when $A = 22°$ and $B = 38°$. Check your answer by computing the sine and cosine of the added angles. Output from the main subprogram.

6.8 Work Exercise 6.6 using two additional subprograms instead of two functions.

6.9 Rework Exercise 4.10 (Chapter 4) but put the factorial calculation in a subprogram and the gamma function calculation in a function.

6.10 The following explicit approximation to the friction factor, f, in turbulent pipe flow is due to Haaland (1983):

$$\frac{1}{\sqrt{f}} = -1.8 \log_{10}\left[\frac{6.9}{Re} + \left(\frac{\varepsilon/D}{3.7}\right)^{1.11} \right].$$

Here, Re is the Reynolds number, ε is the roughness, and D is the pipe diameter.

Write a VBA program to input the Reynolds number, pipe diameter, and rough factor in the main program. Then evaluate f in a function. Output the result from the main program. Note $\log_{10}(x) = 0.43429 * \ln(x)$.

Run your program for $Re = 500,000$ and $\varepsilon/D = 0.004$.

$$\lim_{(x,y)\to(0,0)} \frac{x^4 - y^4}{x^2 + 2y^2}$$

$$\frac{0}{1} = 0$$

$$-\tfrac{1}{2}$$

DNE

7 Roots of Nonlinear Equations: Finding Zero

■ 7.1 What Is a Root?

A root of an equation $f(x) = 0$ is a value of x that satisfies the equation. If

$$f(x) = 2x - 6 = 0,$$

$x = 3$ clearly satisfies the equation and therefore is its root. Some equations, such as polynomials or those involving trigonometric functions, have multiple roots. In such circumstances, finding the one that is physically meaningful may be challenging. Mathematics will not tell us this.

If a function with one independent variable, say, x, is plotted against x, the value of x where the function crosses the x-axis is the root. This is seen in Figure 7.1, the graph of the equation just given where the crossover is at $x = 3$.

The idea that a root is identified where a function crosses, or even touches (a fine point to which we shall return) the independent variable axis, leads to one straightforward method to find a root. Bisection, or the method of interval halving, looks for sign changes that signify the axis has been crossed. It is a reliable method but not very fast. In Figure 7.1, the function is negative for $x < 3$ and positive for $x > 3$. Similarly, $f(1) = -4$ while $f(5) = 4$. Next, we could go to the middle of the region $1 \le x \le 5, x = 3$, where we would find $f(3) = 0$. Usually more steps than this would be required to find the root of the equation. In general we could find the root by systematically looking for sign changes. This method operates, as we have said before

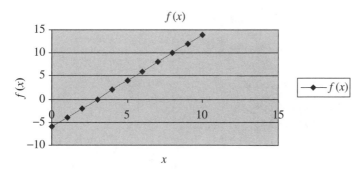

I FIGURE 7.1 Root of a Linear Equation, $2x - 6 = 0$.

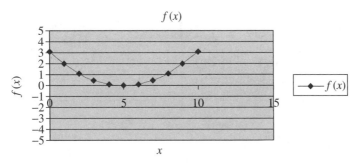

I FIGURE 7.2 Double Root of a Quadratic Equation.

of other numerical methods, by exploiting a fundamental property of the problem—in this case, the sign change that occurs when the axis is crossed.

Although bisection is not as fast as the methods we shall consider in this chapter, it can be used as a preliminary check that a root exists in a certain domain. If the lower limit of that domain is xl and the upper limit is xu, the product $f(xl) * f(xu)$ will be negative if an odd number of roots of $f(x)$ are within that interval. For $f(x) = 2x - 6$, we have $f(1) * f(5) = -4 * 4 = -16$, proof of a sign change in the function between these two locations, and hence a root.

But some functions do not cross the horizontal axis. For example, consider $f(x) = 0.125x^2 - 1.25x + 3.125$ as shown in Figure 7.2. It only touches the horizontal axis and elsewhere is above it. (Actually, there are two roots at $x = 5$.) Hence, looking for a sign change to find a root of this equation would be futile. But this situation is relatively uncommon.

A more common difficulty concerns functions with multiple roots. Consider the function $f(x) = x^2 - 10x + 21$, which is plotted in Figure 7.3. Here we see that the function crosses the horizontal axis twice. But if we evaluate the function at, say, $x = 1$ and $x = 9$, we have positive values at both x's and hence might conclude, incorrectly, that no root exists between these two values. A sign change really means an odd num-

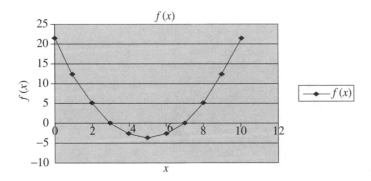

I FIGURE 7.3 Distinct Roots of a Quadratic Equation.

ber of roots exist within an interval; the lack of a sign change can mean no roots or an even number of roots.

■ 7.2 The Newton–Raphson Method

A standard method of finding the root of a nonlinear equation is the Newton–Raphson method. The bisection method might be likened to looking for someone whose exact address you do not know by systematically knocking on doors until you succeed. (Like finding roots, you have to have some idea where to begin.) In contrast, the Newton–Raphson method provides a formula to guide the iteration. Thus, it might be likened to getting information at each house about where the person you are seeking lives. It is an iterative method: that is, a "guess and correct" method, which, like all iterative methods, must begin with a starting value. Later, we shall derive the Newton–Raphson formula and provide a geometrical interpretation. In Section 7.4 we shall consider ways of finding reasonable starting values.

The Newton–Raphson formula for advancing from the current value of x, x_n to the next value of x, x_{n+1} is as follows:

$$x_{n+1} = x_n - \frac{f(x_n)}{f'(x_n)} \tag{7.2.1}$$

where f' is the derivative of f with respect to x.

There are three requirements for successfully applying Newton–Raphson to an equation:

1. The starting value must be reasonably close to the desired root. Otherwise, the solution may diverge or it may converge to a different root for functions with multiple roots.
2. The first derivative obviously must not be 0. If this should happen, the method fails because the next root would be at infinity. You cannot get back from there.
3. The second derivative should not be too large. This is not apparent now but will be demonstrated later.

Application of the Newton–Raphson method requires that we be able to differentiate the function, which may be difficult or at least tedious. To avoid this problem, we shall consider in Section 7.3 a related method, the secant rule, which allows us to avoid actual differentiation by approximating the derivative.

7.2.1 Example

Let us solve the following equation approximately, using Newton–Raphson:

$f(x) = \tan(x) * \sin(x) - 2.15$, x in radians.

Remembering the product rule,

$f'(x) = \sec^2(x) * \sin(x) + \tan(x) * \cos(x)$.

Let $x_0 = 1.47$ (we shall justify this value below). Then,

$f(x_0) = f(1.47) = \tan(1.47) * \sin(1.47) - 2.15 = 7.687$,

and

$f'(1.47) = \sec^2(1.47) * \sin(1.47) + \tan(1.47) * \cos(1.47) = 99.25$.

So,

$$x_1 = 1.47 - \frac{7.687}{99.25} = 1.39.$$

The first five iterations produce the following results:

iter.	x	f(x)	f'(x)	ε_a, %
0	1.47	7.687	99.25	
1	1.39	3.313	32.29	5.74
2	1.29	1.181	13.47	7.75
3	1.20	.2658	8.122	7.50
4	1.17	.0499	6.956	2.56
5	1.167			0.26

$\% \ Error = \left[\dfrac{X_1 - X_0}{X_1} \right] * 100$

That x is converging is evidenced by the bunching of values (smaller values of ε_a), and by the fact that $f(x)$ is getting smaller and smaller as we approach the root. At the root, $f(x)$ will be 0. The value of f', though changing, remains finite.

7.2.2 Correct Solution to the Example

The equation

$$f(x) = \tan(x) * \sin(x) - 2.15 = 0$$

actually can be solved in closed form; hence, the numerical solution really isn't needed.

Note: $\tan(x)$ can be rewritten as $\dfrac{\sin(x)}{\cos(x)}$. Substituting for $\tan(x)$, $f(x)$ can be written as follows:

$$f(x) = \frac{\sin(x)}{\cos(x)} * \sin(x) - 2.15 = 0.$$

And, recalling from trigonometry that

$$\sin(x) * \sin(x) = 1 - \cos^2(x),$$

then

$$f(x) = \frac{(1 - \cos^2(x))}{\cos(x)} - 2.15 = \frac{1}{\cos(x)} - \cos(x) - 2.15 = 0.$$

If we let $z = \cos(x)$ and substitute in the previous equation,

$$f(x) = \frac{1}{z} - z - 2.15 = 0,$$

or, multiplying by z, we obtain $f(x) = 1 - z^2 - 2.15z = 0$.

This quadratic equation is easily solved to yield a positive value of $z = \cos(x) = 0.3932$.

Then $x = \cos^{-1}(0.3932) = 1.167$.

This is the same value we obtained with the Newton–Raphson method, which indicates the correctness of that result. It is to be emphasized that closed-form solutions, when possible, are always preferred over a numerical solution. The equation considered, was chosen to illustrate the ability of the Newton–Raphson method to find a known root.

7.2.3 Derivation of the Newton–Raphson Method

Consider a one-derivative Taylor series for $f(x_{n+1})$ expanded about x_n:

$$f(x_{n+1}) = f(x_n) + f'(x_n) * (x_{n+1} - x_n). \tag{7.2.2}$$

Now assume that x_{n+1} is a root, x_r, of f. (We are labeling this a root but not claiming we know where it is.) Then

$$f(x_{n+1}) = f(x_r) = 0.$$

Equation (7.2.1) thus becomes

$$0 = f(x_n) + f'(x_n) * (x_r - x_n).$$

Solving for x_r we obtain

$$x_r = x_n - \frac{f(x_n)}{f'(x_n)}.$$

Thus we have the Newton–Raphson formula. Based on this, why does the method require iteration at all? Answer: the one-term Taylor series in (7.2.2) omitted all the terms higher than the first derivative. So, in the Newton–Raphson method, we are substituting multiple evaluations of the simple equation rather than evaluating more terms in the Taylor series. This is typical of numerical methods. It is like using the low gear on a bicycle, which is easy to turn but takes multiple turns to cover the same distance as with a higher, but hard-to-turn, gear.

7.2.4 Error Analysis of the Newton–Raphson Method

Let us consider a Taylor series for x_{n+1} through the first-derivative and remainder terms,

$$f(x_{n+1}) = f(x_n) + f'(x_n) * (x_{n+1} - x_n) + \frac{f''(z) * (x_{n+1} - x_n)^2}{2},$$

where $x_n <= z <= x_{n+1}$.

As before, let x_{n+1} be an undetermined root, x_r. Then

$$f(x_{n+1}) = f(x_r) = 0.$$

Now we have

$$0 = f(x_n) + f'(x_n) * (x_r - x_n) + \frac{f''(z) * (x_r - x_n)^2}{2} \tag{7.2.3}$$

Next, consider the Newton–Raphson formula:

$$x_{n+1} = x_n - \frac{f(x_n)}{f'(x_n)}.$$

Multiplying through by $f'(x_n)$ and rearranging terms, we obtain:

$$0 = f(x_n) + f'(x_n) * (x_{n+1} - x_n). \tag{7.2.4}$$

Subtracting (7.2.3) from (7.2.2), we obtain

$$0 = f(x_n) + f'(x_n) * (x_r - x_n) + \frac{f''(z) * (x_r - x_n)^2}{2} - f(x_n)$$
$$- f'(x_n) * (x_{n+1} - x_n).$$

Four terms cancel, leaving us with

$$0 = f'(x_n) * (x_r - x_{n+1}) + \frac{f''(z) * (x_r - x_n)^2}{2}. \tag{7.2.5}$$

The terms in the first parentheses, the difference between x_r and x_{n+1}, is the error at x_{n+1}, E_{n+1}. Likewise, the term in the second pair of parentheses is the current error, E_n.

Making these substitutions and rearranging terms, we obtain

$$E_{n+1} = -\frac{f''(z) * E_n^2}{2 * f'(x_n)}.$$

Several things can be observed.

1. The error change is quadratic—that is, the next error is proportional to the square of the current error. If the current error is small, the next error will be very small. (Of course, if the solution is diverging the next result will be much worse.)

2. Again, we see that if $f'(x)$ is zero, the method will fail.

3. Since the error is proportional to $f''(x)$, that quantity should not be too large if the present error is not to be magnified. This is the last requirement specified earlier. This was not at all obvious before we examined the behavior of the error.

7.2.5 Geometric Interpretation of the Newton–Raphson Method

A geometric interpretation of the Newton–Raphson method follows from Figure 7.4.

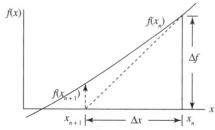

I FIGURE 7.4 Geometric Interpretation of the Newton–Raphson Method.

At $x = x_n$, we can approximate the derivative by constructing the tangent at x_n and extending it down to the x-axis, which yields

$$f'(x_n) \approx \frac{\Delta f}{\Delta x} = \frac{f(x_n) - 0}{x_n - x_{n+1}}.$$

Then $f'(x_n) * (x_n - x_{n+1}) = f(x_n)$.

Solving for x_{n+1}, we obtain the Newton–Raphson formula:

$$x_{n+1} = x_n - \frac{f(x_n)}{f'(x_n)}.$$

Thus, Newton–Raphson operates geometrically by running a tangent line from $f(x_n)$ down to the x-axis to determine x_{n+1}. The slope is again calculated here and the process repeated until convergence is obtained. Of course, if the slope is zero at any x_n, the intersection with the x-axis will be at infinity and the method fails, as we have seen before.

■ 7.3 The Secant Rule

The derivative of $\sin(x) * \tan(x) - 2.15$ was not hard to evaluate, but sometimes evaluation of derivatives is tedious and we would like to avoid it. Even if the differentiation is not difficult, we might want to avoid it to provide a method that can be used by someone with no knowledge of calculus. For example, scientific calculators frequently have a built-in root finder that solves an input equation without requiring that the derivative also be input.

We can avoid actual differentiation if we approximate the derivative and use this approximation instead of the exact formulation in the Newton–Raphson method. This simplifies the process but may require a few more steps to obtain convergence comparable to that achieved with Newton–Raphson. (Remember, in numerical methods you tend to get what you pay for, so to speak. More upfront work—for example, taking a derivative—generally yields a faster or more accurate method.)

Recall the definition of the derivative of $f(x)$ from calculus:

$$f'(x) = \lim_{\Delta x \to 0} \frac{[f(x + \Delta x) - f(x)]}{\Delta x}.$$

Now, we cannot let $\Delta x = 0$ in a numerical procedure, but we can let x and $x + \Delta x$ be two consecutive values in an iteration process; that is, $x = x_{n-1}$ and $x + \Delta x = x_n$.

Then we obtain

$$f'(x_n) = \frac{f(x_n) - f(x_{n-1})}{x_n - x_{n-1}}$$

as an approximation for $f'(x_n)$. If this is substituted for the derivative in the denominator of the Newton–Raphson equation and simplified, we obtain

$$x_{n+1} = x_n - \frac{f(x_n) * (x_n - x_{n-1})}{f(x_n) - f(x_{n-1})}. \tag{7.3.1}$$

Notice that using the secant rule requires using two consecutive values on the right-hand side of the equation. Thus we will need two starting values. This is not more of a problem than with Newton–Raphson, however, for if we have a reasonable starting value for Newton–Raphson, we can simply tweak it a bit to get another starting value. For example, if we again consider $f(x) = \sin(x) * \tan(x) - 2.15$, we might use $x_0 = 1.45$ and $x_1 = 1.5$, which bracket the value of 1.47 that we had used to start the Newton–Raphson method. Ultimate convergence is not affected by which value we call x_0 and which x_1, though the actual numerical history of the solution will depend on how we number the starting values.

Let us now apply the secant rule to this problem.

iter	x	f(x)	ε_a, %
0	1.45	6.028	
1	1.50	11.92	
2	1.399	3.522	7.22
3	1.356	2.336	3.17
4	1.273	0.961	6.52
....	....	....	....
7	1.168	.08862	0.856
8	1.167		0.771

Seven iterations beyond x_1 were required to get the same answer that we obtained with Newton–Raphson in five; however, no differentiation was required.

7.3.1 General Numerical Derivative

We may generalize the procedure for finding a derivative numerically. Consider again the definition of a derivative:

$$f'(x) = \frac{\lim}{\Delta x \to 0} \frac{\Delta f}{\Delta x}.$$

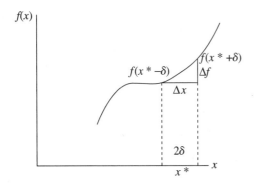

I FIGURE 7.5 General Numerical Derivative.

At some point x^* we could express Δf as $f(x^* + \delta) - f(x^* - \delta)$ and Δx as $2 * \delta$. Then,

$$f'(x) \approx \frac{f(x^* + \delta) - f(x^* - \delta)}{2 * \delta}.$$

This relationship is depicted in Figure 7.5.

We know that δ cannot approach 0 in the limit, as in the definition of a derivative, but it should be small, say 0.0001. We are actually constructing a secant line between two points near x^* and assuming that the tangent through x^* is parallel to this secant.

Let us consider again the function $f(x) = \tan(x) * \sin(x) - 2.15$. If $x^* = 1.47$ and $\delta = 0.0001$, $f(1.4701) = 7.6971$ and $f(1.4699) = 7.6773$. Then

$$f'(x) \approx \frac{7.6971 - 7.6773}{0.0002} = 99.00,$$

which is close to the correct value of 99.25 as we determined in Section 7.2.1.

■ 7.4 Estimating Starting Values

Iterative methods of finding roots require that the process begin with a starting value. For nonlinear equations, the choice of a reasonable starting value is essential. Otherwise, the process may not converge.

Fortunately, in practical problems we often (and should) have a rough idea what the answer is. For example, if we are doing an HVAC problem we may assume that the room temperatures will be about 70°F. In a circuits problem, we may know that the voltage is about 120. Thus, we can use our understanding of the physical problem to estimate the starting value and then use a root-finding technique to find the precise value we are seeking. In this respect, our task is easier than it is for mathematicians who may be looking from $-\infty$ to $+\infty$ for a root.

But if we lack this information, we may begin by first approximating the actual equation with a simplified form and solving that. The exact answer to our approximate equation can provide an excellent starting value for solving the exact equation. We shall now consider several ways to attempt to estimate a good starting value. No guarantees of success are offered.

7.4.1 Series: The Example Starting Value

Consider the equation that we solved previously with both the Newton–Raphson root-finding method and the secant rule:

$$f(x) = \tan(x) * \sin(x) - 2.15 = 0.$$

Using a Maclaurin series (refer to Chapter 3), we may approximate $\tan(x)$ as $x + \dfrac{x^3}{3}$ and $\sin(x)$ as x. Then our equation becomes

$$\left(x + \frac{x^3}{3}\right) * x - 2.15 = 0.$$

Expanding this, we obtain

$$x^2 + \frac{x^4}{3} - 2.15 = 0.$$

The x^4 term still poses a problem so let us drop it, too. (There is no point in approximating an equation with something that is still difficult to solve, just to find a starting value.) So the equation becomes

$$x^2 = 2.15 \text{ or } x = \sqrt{2.15} = 1.47,$$

which is the value we successfully used previously. If we had included the fourth-order term, we might have let $x^2 = z$ and solved the resulting quadratic equation $z + \dfrac{z^2}{3} - 2.15$ for z, yielding $z = 1.45$. Then $x = \sqrt{z} = 1.20$, which is quite close to the actual root at $x = 1.167$. But we need not work this hard just to obtain a starting value.

7.4.2 Equations with Power Terms

To better illustrate the discussion, let us examine an application, such as the following problem that occurred in a fluid mechanics class:

$$\Delta P - 5.4v^{1.8} = 0$$

where ΔP is the pressure drop and v is the velocity. If v is known, solving this equation for ΔP is trivial. But in the actual problem the pressure drop, ΔP, was given as 1440 lb/ft^2. The problem was to find the corresponding velocity. If the power, 1.8, is replaced by 2 (which is not a big change), the approximate equation is easy to solve:

$$\Delta P = 5.4v^2 \text{ or } \sqrt{(1440/5.4)} = v = 16.3.$$

This would provide a good starting value of the real equation containing the $v^{1.8}$ term. The root of that equation is 22.3.

7.4.3 A Polynomial Equation

If we have a polynomial equation of the form

$$ax^n + bx^{n-1} + cx^{n-2} + \ldots + zx + const = 0,$$

we may estimate the largest root by retaining only the highest power term and the constant: for example $ax^n + const$ and solving for x. Likewise, we may estimate the smallest root by retaining only the linear term and the constant, $zx + const = 0$ and solving for x. These values may serve as initial values for solving the equation with an iterative method. If the root is less than 1.0, the methods for the largest and smallest roots will be reversed.

Consider the cubic equation

$$x^3 - 2.55x^2 + 0.900x - 0.0400 = 0.$$

Retaining only the cubic term and the constant yields

$$x^3 = 0.0400,$$

which can be solved to obtain $x_1 = 0.342$. In fact, a root exists at $x = 0.360$.

Likewise, the smallest root may be estimated by retaining only $0.900x - 0.0400$, which may be solved to obtain $x_s = 0.044$. There is an actual root at $x = 0.052$.

These are good results that might not be typical. In the preceding cubic equation for example, if the constant were +0.0400, this procedure would yield, for the estimated large root, a negative result and a pair of complex values. The full equation actually has three real roots, so this time the estimate did not work. Thus, this way of estimating, like the others, might fail. When that happens, we must try another approach. We need multiple tools in our toolbox.

7.4.4 Two Simultaneous Nonlinear Equations

Consider the pair of equations

$$f(x,y) = x * \sin(y) - e^x + 3.74 = 0, \text{ and} \tag{7.4.1}$$
$$g(x,y) = y * \cos(x) + e^y - 1.24 = 0; \tag{7.4.2}$$

where x and y are in radians.

If we make the assumption that x and y are small, we may use short Maclaurin series approximations for $\cos(x)$ as 1 and $e^y = 1 + y$ in (7.4.2).

Then (7.4.2) becomes

$$g(x,y) = y * 1 + (1 + y) - 1.24 = 0,$$

which is easily solved for $y = 0.120$.

Using this y and approximating e^x as $1 + x$ in (7.4.1) yields

$$f(x,y) = x * 0.120 - (1 + x) + 3.74 = 0,$$

which yields $x = 3.11$.

Though y is a small value, this estimate of x is not small. Nevertheless, we will try it shortly as it is only a starting value.[1]

■ 7.5 Two-Equation Newton–Raphson Method

Now let us consider two equations with two unknowns, $f(x,y) = 0$ and $g(x,y) = 0$.

To obtain the next values of x and y, the partial derivatives of f and g with respect to x and y (thus four partials in all) must be computed and used in the following formulas. These formulas are much more complex than that in the single-variable Newton–Raphson method. Hence, this procedure is much more tedious, and it is impractical for solution by hand.

$$x_{n+1} = x_n - \frac{\left(f\dfrac{\partial g}{\partial y} - g\dfrac{\partial f}{\partial y}\right)}{\dfrac{\partial f}{\partial x}\dfrac{\partial g}{\partial y} - \dfrac{\partial f}{\partial y}\dfrac{\partial g}{\partial x}} \tag{7.5.1 a}$$

---

[1] I am grateful to UT-Chattanooga Engineering Professor Emeritus J. Eric Schonblom for suggesting the problem of two simultaneous nonlinear equations in Section 7.4.4 and the technique in Section 7.4.3.

and

$$y_{n+1} = y_n - \frac{\left(g\dfrac{\partial f}{\partial x} - f\dfrac{\partial g}{\partial x}\right)}{\dfrac{\partial f}{\partial x}\dfrac{\partial g}{\partial y} - \dfrac{\partial f}{\partial y}\dfrac{\partial g}{\partial x}} \qquad (7.5.1b)$$

Using these equations requires evaluating both f and g and the four partial derivatives at each iteration. Applied to Equations (7.4.1) and (7.4.2) and using the starting values just found in Section 7.4.4, the first six iterations yield the following results:

Iteration	x	y	f	g
0	3.11	0.120	−18.3	−0.232
1	2.53	1.92	−6.45	3.99
2	2.03	1.16	−2.03	1.43
3	1.65	0.489	−0.686	0.357
4	1.41	0.185	−0.0976	0.00397
5	1.39	0.185	−0.00127	−0.0000209
6	1.38	0.185	$-2 * 10^{-7}$	$-1 * 10^{-8}$

The very small values of f and g after six iterations indicate that the method has essentially converged. Notice that this occurred even though the initial guess for x was neither small nor close to its final value.

■ 7.6 General Nonlinear Equation Sets

Press et al. (1989) state there are no reliable methods for finding the roots of a set of nonlinear equations. However, the following procedure, based on the Newton–Raphson method, can be tried, although convergence is not guaranteed.

Let $f_i(x_i) = 0$ be a set of n equations in the n unknowns, x_i, $i = 1$ to n. Let the current values of the x's be perturbed by an amount, δ_i. Thus, $x_i^{k+1} = x_i^k + \delta_i$ for all i where the superscript indicates the iteration.

If we write a Taylor series for the f_i through only the first derivative terms in all i, we obtain

$$f_i(x_i + \delta_i) = f(x_i) + \sum_{j=1}^{j=n} \frac{\partial f_i}{\partial x_j} * \delta_j,$$

where the partials are evaluated at the current x_i. Optimistically assuming that all the $(x_i + \delta_i)$ satisfy the f_i on the left-hand side, we obtain

$$0 = f_i(x_i) + \sum_{j=1}^{j=n} \frac{\partial f_i}{\partial x_j} * \delta_j, \tag{7.6.1}$$

a set of linear equations in the δ_j. These may be solved using, say, Gauss elimination (as will be explained in Chapter 9). When the δ's are added to the current x's, the new values of x_i are obtained. This process may be repeated until (and if, remember no guarantees) convergence is obtained.

For the $f(x,y)$ and $g(x,y)$ specified in Equations (7.4.1) and (7.4.2), the equations in (7.6.1) become

$$0 = f(x,y) + \frac{\delta f}{\delta x}\Big|_y \delta_x + \frac{\delta f}{\delta y}\Big|_x \delta_y, \tag{7.6.2a}$$

and

$$0 = g(x,y) + \frac{\delta g}{\delta x}\Big|_y \delta_x + \frac{\delta g}{\delta y}\Big|_x \delta_y, \tag{7.6.2b}$$

where $\dfrac{\delta f}{\delta x}\Big|_y = \sin(y) - e^x$ and $\dfrac{\delta f}{\delta y}\Big|_x = x * \cos(y)$,

and

$$\frac{\delta g}{\delta x}\Big|_y = -y * \sin(x) \qquad \text{and} \qquad \frac{\delta g}{\delta y}\Big|_x = \cos(x) + e^y.$$

We can solve equations (7.6.2a) and (7.6.2b) for δ_x and δ_y. Then,

$$x^{k+1} = x^k + \delta_x \text{ and } y^{k+1} = y^k + \delta_y,$$

and we repeat the process until convergence is obtained (if it converges).

The partials, $\dfrac{\partial f}{\partial x}, \dfrac{\partial f}{\partial y}, \dfrac{\partial g}{\partial x}$, and $\dfrac{\partial g}{\partial y}$, may be approximated numerically by perturbing f and g by x and y and computing the change in f and g with those changes: for example,

$$\frac{\partial f}{\partial x} = \frac{f(x + \Delta x) - f(x)}{\Delta x} \qquad \text{and} \qquad \frac{\partial f}{\partial y} = \frac{f(y + \Delta y) - f(y)}{\Delta y}.$$

And we proceed similarly for g.

■ 7.7 Multiple Equations with the Secant Rule

Another approach to attempt to find the roots of a set of nonlinear equations (again, no guarantees of success) is to advance only one root in each equation using the secant rule each iteration.

Now we will consider that $f(x,y) = 0$ and $g(x,y) = 0$. We will solve f for x and g for y, in turn, remembering that we are dealing now with partial derivatives of f and g. The choice of which to solve for x and which for y is arbitrary. Thus,

$$x_{n+1} = x_n - \frac{(f(x_n, y_n) * (x_n - x_{n-1}))}{(f(x_n, y_n) - f(x_{n-1}, y_n))}. \tag{7.7.1a}$$

Note that y_n is fixed and that

$$y_{n+1} = y_n - \frac{(g(x_{n+1}, y_n) * (y_n - y_{n-1}))}{(g(x_{n+1}, y_n) - g(x_{n+1}, y_{n-1}))}, \tag{7.7.1b}$$

where x_{n+1} is used immediately and fixed for the value of x.

Fixing y in the x equation and x in the y equation is correct because we are approximating partial derivatives. Using the latest value of x in the y equation is reasonable. We shall return to this point in Chapter 9 when we consider the Gauss–Seidel method of solving sets of linear equations.

After the new x and y are found, the approximate relative errors of the change may be computed and each compared with the tolerance. If either is too large, a new step must be performed. This process must be repeated until convergence is obtained, or *it should be* discontinued if a lack of convergence is demonstrated. As in all nonlinear iteration schemes, the starting values may be very important in pursuing convergence. This procedure could be extended to more equations and unknowns.

■ 7.8 Flowchart: Two Equations with the Secant Rule

The flowchart in Figure 7.6 applies the just-described scheme to a set of two equations. Note that, for generality, the program defines the two equations in functions. Thus, all that needs to be changed within the program to apply it to a different equation pair is in the definitions of these two functions. This is an example of the generality that we can achieve in programming through the use of functions.

Results from the application of this program to the two equations in Section 7.4.4 follow. The program converged in 11 iterations beyond the starting values to an accuracy of better than seven significant figures (`errx` and `erry` as percentages, each less than $0.5 * 10^{-5}$).

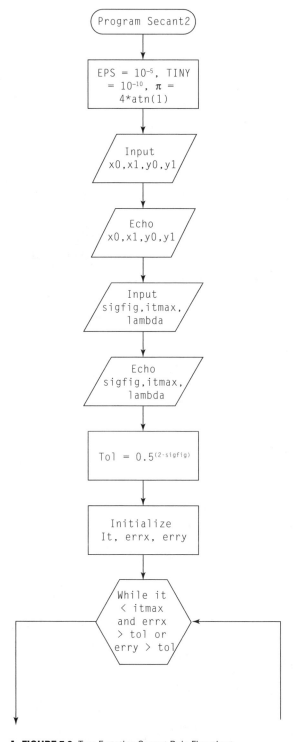

❙ FIGURE 7.6 Two-Equation Secant Rule Flowchart.

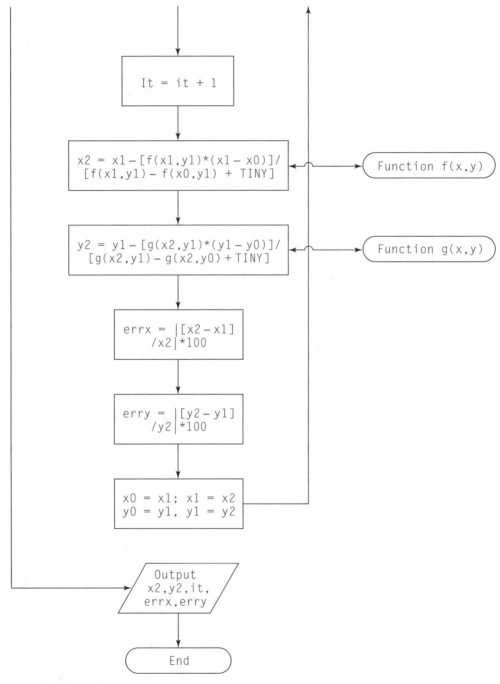

I FIGURE 7.6 *(Continued)*

> **TIP:**
>
> Notice also the use of the small $(1 * 10^{-10})$ variable TINY in the denominator of the x and y equations. This "trick" is employed to prevent division by zero in the event that the value of f or g is identical for both sets of arguments. This may happen, especially at the beginning of the iteration process. If f and g are identical, computation would cease without the insertion of this small value; on the other hand, if they are not identical, its use has virtually no impact.

it	x	y	errx	erry
−1	3.10	0.20		
0	3.00	0.10		
1	2.23679	0.495329	34.121	79.8114
2	1.90084	0.262144	17.6735	88.953
3	1.58123	0.222939	20.2125	17.5858
4	1.44115	0.195467	9.71998	14.0541
5	1.39378	0.188157	3.39864	3.88521
6	1.38633	0.187057	0.537639	0.588381
7	1.38572	0.186972	0.0439779	0.0452575
8	1.38569	0.186968	0.00236698	0.00234317
9	1.38569	0.186967	0.000113727	0.000111556
10	1.38569	0.186967	$5.37863 * 10^{-6}$	$5.27191 10^{-6}$
11	1.38569	0.186967	$2.54091 * 10^{-7}$	$2.49039 * 10^{-7}$

■ 7.9 The Excel Solver

The Excel Solver provides an effective means of solving one or more nonlinear equations. Like all iterative procedures, it requires starting value(s) that the user must supply. The Solver is activated from the Tools pull-down menu. If it does not appear there, it will be necessary to activate it by selecting "Add-Ins", and then clicking the "Solver Add-in" box.

To use the Solver, we must enter the starting value of the unknown(s) into the spreadsheet and then define the equation(s) in terms of the unknown(s) referenced to the cell(s) containing the starting value(s). Labeling these variables in adjacent cells is useful. Then the Solver is used to iterate on that starting value(s) in order to set the equation value(s) equal to 0.

7.9.1 One Equation

Suppose we want to solve the example equation considered in Section 7.2.1,

$$f(x) = \tan(x) * \sin(x) - 2.15 = 0,$$

where x is in radians.

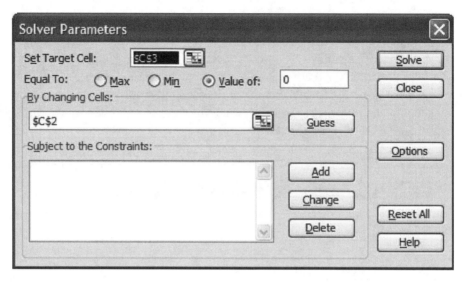

I FIGURE 7.7 Excel Solver Dialog Box.

We will place a label, $x=$, arbitrarily in cell B2 and a starting value in cell C2. For simplicity, enter 0 in cell C2. (If x were in the denominator of any term in the equation, this choice would not have worked because we would have had a division by zero.)

In cell B3 we shall type in the label, $f(x)=$, and we shall enter the equation in cell C3 in terms of the x value in cell C2. Thus, in C3 we shall enter

$$= \tan(C2) * \sin(C2) - 2.15.$$

We must not forget the initial equals sign $(=)$. Otherwise, we would simply have a text form of the equation.

When the Solver is activated, a Solver Parameters dialog box (Figure 7.7) will display. The Target Cell in our case is cell C3 (the value of the function), which should become 0 when the equation is solved. We can select C3 by either typing this in the "Select Target Cell" window or clicking the C3 cell itself. In the next line of the dialog box, we click the "Value of" radio button and enter 0, the value we want for our correctly solved equation. Then we tell the Solver to attempt this solution by changing the value of the unknown, x, which is in cell C2. We can either type the cell location in the "By Changing Cells" window or click the C2 cell itself.

Then click the "Solve" button in the upper right of the dialog box to obtain a solution. Almost immediately after clicking Solve, we see the solution in cell C2 and the value of the function at that x in cell C3. After solution the spreadsheet will show the labels in B2 and B3, and the resulting values in C2 and C3:

(B2): x = (C2): 1.66682

(B3): $f(x)$ = (C3): $-7.5e - 07$

The value of $f(x)$ is not zero, but it is close, consistent with a single-precision calculation. If for some reason a solution is not obtained, a gray dialog box will inform us.

In addition, Excel lets us apply constraints to the solution within the dialog box. For example, we could specify a solution with $x > 0$, though such a solution might not actually exist.

7.9.2 Multiple Equations

Multiple equations are solved similarly. Each equation will equal 0 at its root. But we must be clever, because though we have more than one unknown, we can set only one value for the Target cell in the Solver Parameters dialog box. We will again define each equation in terms of the cell locations of the unknowns. We could then require that the sum of the solutions to the functions equals 0. This is true but it is not enough. The sum of the function values could be zero or close to zero even if none of the function values were. How? If the function values included both positive and negative results, their values could cancel each other. To get around the possibility of negative cancellation, we will square each function value and then require the sum of the squares to be close to or equal to 0. Squaring the function values will make them all positive and hence avoid cancellation of positive and negative results.

Consider the pair of equations introduced in Section 7.4.4:

$$f(x,y) = x * \sin(y) - e^x + 3.74 = 0 \quad \text{and}$$
$$g(x,y) = y * \cos(x) + e^y - 1.24 = 0,$$

where x and y are in radians.

We shall put labels for x and y in cells B2 and B3 and their starting values in cells C2 and C3. We shall label $f(x,y)$ in cell B4 and define it in C4 as = C2 * sin(C3) − exp(C2) + 3.74. $g(x,y)$ will be labeled in B5 and defined in C5 as = C3 * cos(C2) + exp(C3) − 1.24. The sum of f^2 and g^2 will be labeled in B6 and defined as = C4^2 + C5^2 in C6.

Next we will run the Solver. We will try to set the target cell value in C6 (the sum of f^2 and g^2) equal to zero by changing the values of x and y in the cell range C2:C3. After solution, the spreadsheet results will be as follows:

x =	1.385630698
y =	0.187311208
f(x,y) =	0.000682886
g(x,y) =	0.000488285
f^2 + g^2 =	7.04755E–07

■ 7.10 Chapter 7 Exercises

7.1 Consider the equation $f(x) = \tan(x) - \sin(x) - 2.75 = 0$; x is in radians.

   a. By evaluating $f(x)$ at $x = 0.5$ and $x = 1.5$ and comparing the signs, show that a root exists between these two points.

   b. Write a Maclaurin series for $\sin(x)$ and $\tan(x)$ through the third derivative terms. Substitute these series in the original equation. Solve the resulting simplified algebraic equation in closed form.

   c. Use your answer to **b.** to start a Newton–Raphson solution to the **original** equation. Perform 3 iterations and compute the ε_a's for each iteration.

7.2 Consider the equation specified in Exercise 7.1. Apply the secant rule to this equation for 3 iterations beyond the starting values. Compute the ε_a's for each iteration.

   a. Use x = 2.0 and x = 4.0 as starting values.

   b. Use x = 1.0 and x = 1.5 as starting values.

7.3 Consider the equation $f(x) = \tan(x) - \dfrac{1}{0.720 * x} = 0$.

   a. Write a Maclaurin series for $\tan(x)$ through the second non-zero term. Substitute this for the $\tan(x)$ term in $f(x)$. Solve the resulting simplified algebraic equation in closed form.

   b. Use your answer to **a.** to start a Newton–Raphson solution to the original equation. Perform 3 iterations and compute the ε_a's for each iteration.

7.4 Consider the equation given in Exercise 7.3. Estimate two starting values (you are on your own) and then perform 3 iterations beyond the starting values with the secant rule. Compute the ε_a's for each iteration.

7.5 Consider the equation $f(x) = e^x + \dfrac{\tan(x)}{x} - 3 = 0$.

   a. By evaluating $f(x)$ at $x = 0.1$ and $x = 1.0$ and comparing the signs, show that a root exists between these two points.

   b. Use your answer to **a.** to start a Newton–Raphson solution to the equation. Perform three iterations and compute the ε_a's for each iteration.

7.6 Consider the equation in Exercise 7.5.

   a. Write a Maclaurin series for $\tan(x)$ through the second non-zero term. Substitute for $\tan(x)$ in $f(x)$ and solve the resulting simplified algebraic equation in closed form.

b. Use your answer to **a.** as the starting value and perform three iterations to solve the *original* equation for a positive value with the secant rule. Compute the ε_a's for each iteration.

7.7 Solve the following equation for T using the Newton–Raphson method. Continue until the change in T is less than $1°$. The equation came from a heat transfer problem involving both convection and surface radiation.

$$4000 = 125 * (T - 305) + 5.67 \times 10^{-8} * 9.3 * (T^4 - 305^4)$$

a. Estimate a starting value.

b. Use your answer to **a.** to start a Newton-Raphson solution to the equation. Perform three iterations and compute the ε_a's for each iteration.

7.8 Consider the equation given in Exercise 7.7.

a. Show a root exists between $T = 305$ and $T = 500$ °F

b. Perform four iterations with the secant rule to find T. Compute the approximate relative errors

7.9 Consider the equation $f(x) = x^3 - 6x^2 + 9x - 4 = 0$.

a. Estimate the smallest root for this equation.

b. Use your answer to **a.** as the starting value to perform three iterations with the Newton–Raphson method and compute the ε_a's for each iteration.

7.10 Consider the equation in Exercise 7.9.

a. Estimate the largest root of this equation.

b. Use your answer to a to perform three iterations with the secant rule and compute the ε_a's for each iteration.

7.11 Consider the equation $f(x) = \pi/4 - x*(1 - x^2) - \sin^{-1}(x) = 0$.

a. Show by a simple test that a root exists between $x = 0$ and $x = 1$.

b. Perform three iterations beyond the starting values using the secant rule.

c. Estimate the derivative at your answer to **b.** Compare to the correct value.

7.12 Consider the equation in Exercise 7.11.

a. Use the result from Exercise 3.6 that $\sin^{-1}(x) \approx x + \dfrac{x^3}{6}$ to approximate the arcsin in the equation. Solve the resulting approximate equation for x.

b. Use your answer to **a.** as the starting value to perform three iterations with Newton–Raphson and compute the ε_a's for each iteration.

7.13 Consider the equation $(\sin(x))^2 - x + 0.3 = 0$, x in radians.

a. Substitute a Maclaurin series for $\sin(x)$ through the second derivative term in the above equation. Solve the resulting simplified algebraic equation in closed form.

b. Use your answer to **a.** to start a Newton–Raphson solution to the original equation. Perform three iterations and compute the ε_a's for each iteration.

7.14 Solve the problem in Exercise 7.13 with the secant rule. Perform three iterations beyond the starting values and compute the ε_a's for each iteration. Use $x_0 = 1.0$ and $x_1 = 1.4$.

7.15 Consider the equation $f(x) = 7.94\,x - \sqrt{\dfrac{0.335}{1 + 0.2x^2}} = 0$.

a. Estimate a starting value.

b. Use your answer to **a.** to start a Newton–Raphson solution to the original equation. Perform three iterations and compute the ε_a's for each iteration.

c. Solve the equation with the secant rule. Perform three iterations and compute the ε_a's for each iteration.

7.16 The following equation is used to calculate the heat transfer in a boiler given the surface area, A, the overall heat transfer coefficient, U, the heating fluid inlet temperature, T_1, the heating fluid exit temperature, T_2, and the temperature of the boiling fluid, T_s.

$$Q = U\!*\!A\,\frac{T_1 - T_2}{\ln\left\{\dfrac{(T_1 - T_s)}{T_2 - T_s}\right\}}\,.$$

If $T_1 = 600\ {}^\circ F$, $T_s = 535\ {}^\circ F$, and $\dfrac{Q}{UA} = 34.1\ {}^\circ F$, determine T_2.

Use Newton–Raphson and perform three iterations

7.17 Use the secant rule to solve the problem in Exercise 7.16. Perform three iterations beyond the starting values.

7.18 The Colebrook formula (Colebrook, 1939) can be used to determine the friction factor, f, for turbulent flow through a rough pipe as a function of the Reynolds number, Re, and the relative roughness, $\dfrac{\varepsilon}{D}$, where ε is the equivalent roughness and D is the pipe diameter.

$$\frac{1}{\sqrt{f}} = -2.0\log_{10}\left(\frac{\varepsilon/D}{3.7} + \frac{2.51}{\mathrm{Re}\sqrt{f}}\right)\,.$$

last problem

If Re $= 500,000$ and $\dfrac{\varepsilon}{D} = 0.004$, determine f using Newton–Raphson. Assume $f_0 = 0.20$, a "universal" starting value for such problems.

7.19 Solve Exercise 7.16 using the Excel Solver.

7.20 Heat transfer in a certain process involving radiation is described by the equations

$$Q = \frac{T_2 - T_3}{85 + R_3} \text{ and } R_3 = \frac{T_2 - T_3}{1.07 \times 10^{-11} * (T_2{}^4 - T_3{}^4)}$$

If $Q = 3.71$ BTU/h and $T_2 = 1000°$R, calculate T_3 using the Excel Solver.

7.21 Solve the following pair of equations for x and y using the Excel Solver. Turn in spreadsheets showing your answers and also the formulas used. Formulas may be displayed by selecting "Tools/Option" and clicking on the "Formulas" box in the lower right corner of the View tab.

$$x^2 - x + y = 0.5$$
$$x^2 - y - 5xy = 0.$$

7.22 Write a secant rule program using the flowchart in Section 7.8. First, use your program to verify the solutions to Equations 7.7 and 7.8, which were obtained using the two-equation Newton–Raphson equations, 7.9a and 7.9b. Output your roots, the approximate relative errors, the number of iterations required, and the functions evaluated at the roots.

7.23 Write a secant rule program using the flowchart in Section 7.8. Apply your program to the following pair of equations. Estimate your own starting values. Output your roots, the approximate relative errors, the number of iterations required, and the functions evaluated at the roots.

$$f(x,y) = 4 * x_1{}^3 - (3 * x_2)^{0.5} - 20 = 0$$
$$g(x,y) = x_1 * x_2{}^2 + \frac{2}{x_1} - 50 = 0.$$

8 Ordinary Differential Equations: Take One Step Forward

■ 8.1 The Basic Idea

Ordinary differential equations are ubiquitous in engineering and science. Many problems in these fields deal with change, and the equations of change are differential in nature. Ordinary differential equations have only one independent variable—say, time. Many can be solved in closed form, but more cannot. Fortunately, the numerical methods available for solving ordinary differential equations are excellent. Partial differential equations have more than one independent variable and are therefore harder to solve, in closed form or numerically. The numerical solution of partial differential equations will be introduced in Chapter 12.

Consider the mathematical definition of the time derivative of V, where V might be the velocity:

$$\frac{dV}{dt} = \lim_{\Delta t \to 0} \frac{\Delta V}{\Delta t}.$$

Clearly, we cannot let the denominator equal 0 in a calculation. But we can approximate the derivative by taking a small but finite value of Δt. "How small" is a good question, analogous to, but more difficult than, deciding how many intervals to take in performing numerical integration. Some methods of numerically solving ordinary differential equations provide error estimates; most do not. We shall consider evaluating the step size in more detail later.

For purposes of illustration, let

$$\frac{dV}{dt} \approx \frac{\Delta V}{\Delta t} \text{ where } \Delta t \text{ is small,}$$

and let $\dfrac{dV}{dt}$ be represented by $f(V,t)$.

Thus, $\dfrac{dV}{dt} \approx \dfrac{\Delta V}{\Delta t} = f(V,t).$

Multiplying by Δt, we obtain

$$\Delta V = f(V,t) * \Delta t$$

or

$$V(t + \Delta t) - V(t) = f(V,t) * \Delta t.$$

Then $V(t + \Delta t) = V(t) + f(V,t) * \Delta t$

or

$$V(t + \Delta t) = V(t) + \left.\frac{dV}{dt}\right|_t * \Delta t. \tag{8.1.1}$$

As we can see, the new value of V at $t + \Delta t$ equals its previous value at t plus the product of its slope and the intervening time interval. Intuitively, the smaller the step, the better the approximation. This is the basic idea in solving ordinary differential equations numerically. Figure 8.1 illustrates that we have used a straight-line projection to approximate the behavior of $V(t)$ over one time step.

A subtle but important point to consider is, where do we evaluate the derivative? For now we shall evaluate the derivative at the beginning of the step. More elaborate methods, which we shall consider shortly, use combinations of the derivative evaluated at different points within the interval. Some methods use the derivative at the end of the step.

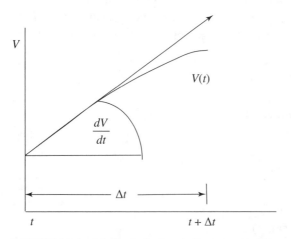

| FIGURE 8.1 Straight-line Projection of a Dependent Variable.

To evaluate the derivative at the beginning of the step in Equation (8.1.1) is Euler's method. It is easy to understand and use, although not especially accurate. But it is a start.

By the way, the literature on ordinary differential equation solvers can be confusing. There are many methods, some very similar to each other, and the naming of them is not always consistent. For example, (1) the *Euler method*, (2) the *improved Euler method*, and (3) the *modified Euler method* are similar names and surely easy to confuse. Moreover, some texts refer to the improved Euler method as *Heun's method*. So be careful in consulting references with regard to the names of the methods.

Let us look at Euler's method again:

$$V(t + \Delta t) = V(t) + f(V,t)|_t * \Delta t.$$

This is just a one-term Taylor series for $V(t + \Delta t)$ in terms of $V(t)$. Hence, it is as accurate such as a Taylor series and therefore is referred to as a first-order method. Later, we shall consider second-order methods, a fourth-order method in detail, and, briefly, a fifth-order method.

■ 8.2 Example: Vehicle Velocity

If we consider an object such as an automobile changing speed, its net acceleration would depend on the balance between the driving force and the resistance. From Newton's second law, the equation of motion would be

$$F - D = m\frac{dV}{dt}.$$

Here F is the net forward force on the wheels, and D, the drag force, is the wind resistance. Wind resistance is approximately proportional to the velocity squared, so we can write D as cV^2 where c is a problem-specific constant that takes into account air density and the cross-sectional area of the vehicle. Hence,

$$F - cV^2 = m\frac{dV}{dt}.$$

Thus, we have a first-order (first derivative) ordinary (not partial) nonlinear (V is squared) differential equation. For an automobile of moderate size and power, we might assume the values as

$$F = 880 \text{ lb}$$
$$m = 3220 \text{ lb}_m = \frac{3220}{32.2} = 100 \text{ slugs}$$
$$c = 0.137.$$

Then our equation becomes

$$880 - 0.137V^2 = 100\frac{dV}{dt},$$

or

$$8.80 - 1.37 \times 10^{-3} V^2 = \frac{dV}{dt} \tag{8.2.1}$$

Although this equation is nonlinear, it can be solved in closed form (many nonlinear equations cannot be solved). If we assume $V(0) = 0$ as the initial condition (i.e., the automobile starts from rest), we obtain the nonobvious closed-form solution

$$V(t) = 80.15 * \tanh(0.1098t). \tag{8.2.2}$$

We shall compare our numerical solution to this result.

Initial conditions (or the equivalent) must always be provided if we are to solve differential equations numerically. Without them, we would not have enough values to proceed. This requirement is equivalent to requiring that initial conditions be provided in closed-form solutions as well, if the solution is not to have one or more undetermined constants—for example, C_1, C_2. We cannot have undetermined C's in a numerical solution.

■ 8.3 Application of the Euler Method

Given (8.2.1), $\dfrac{dV}{dt} = 8.80 - 1.37 \times 10^{-3} V^2$ and $V(0) = 0$.

We shall apply Euler's method to Equation (8.2.1) with $V(0) = 0$, arbitrarily choosing a time step of 1.0 seconds.

At $t = 0$,
$$\frac{dV}{dt} = 8.80 - 1.37 \times 10^{-3} * 0^2 = 8.80.$$

Then, from Equation (8.1.1),

$$V(1.0) = V(0) + \frac{dV}{dt}\Delta t = 0 + 8.80 * 1.0 = 8.80 \text{ ft/s.}$$

Using Equation (8.2.2), we find the correct value is

$80.15(\tanh(0.1098 * 1.0)) = 8.77$.

Thus we have a true relative error,

$$\varepsilon_T = \left| \frac{8.77 - 8.80}{8.77} \right| * 100 = 0.342\%,$$

which is not bad but certainly not satisfying.

Let us do another step. At $t = 1.0$,

$$\frac{dV}{dt} = 8.80 - 1.37 \times 10^{-3} * 8.80^2 = 8.69.$$

Then

$$V(2.0) = V(1.0) + \frac{dV}{dt} \Delta t = 8.8 + 8.69 * 1.0 = 17.5 \text{ ft/s}.$$

Again, from (8.2.2) the correct value is 17.3 ft/s.

Thus we now have a true relative error,

$$\varepsilon_T = \left| \frac{17.3 - 17.5}{17.3} \right| * 100 = 1.16\%.$$

This error is larger than the previous error. If the error were always to increase, we would have an unstable solution, and eventually the solution would produce obviously nonsensical results. This is not the situation here. We shall investigate instability briefly in Section 8.11.

In both cases, the Euler result was greater than the correct solution. No generalization should be made from this. However, the excess is reasonable in this problem since the velocity is continuously increasing and so is the drag. Therefore, the acceleration term evaluated at the beginning of the step is larger than it will be later in the same step. Since the acceleration used in the Euler method is too large over the step, the velocity calculated also will be too large.

■ 8.4 Other Euler Methods

If we were to reduce the step size, Δt, to 0.25, one-quarter its previous value, the error would be reduced. This could be done. But it would be more effective to use a method of higher order—that is, one that agrees with more terms of a Taylor series. The improved

Euler method (also called Heun's method) is a second-order method that is an extension of the basic Euler method.

Suppose at the end of the Euler step we again evaluated the derivative at $t + \Delta t$ before completing the step; that is, we calculated

$$\frac{dV}{dt}\Big|_{t+\Delta t}.$$

This would be based on the Euler value,

$$V(t + \Delta t) = V(t) + \frac{dV}{dt}\Big|_t.$$

We could then average this derivative at the end of the step, with the derivative calculated at the beginning of the step, to get a better representation of the derivative over the entire step. It would not be perfect; the derivative at the end of the step depends on the Euler value at the end of the step. But this average is a better estimate than a single value at the beginning of the step. It is like a business making a projection for the year based on earnings in January and anticipated earnings at the end of December, rather than relying only on the known earnings in January.

The resulting formula for one step of the improved Euler method is

$$V(t + \Delta t) = V(t) + \frac{1}{2}\left[\frac{dV}{dt}\Big|_t + \frac{dV}{dt}\Big|_{t+\Delta t}\right]\Delta t.$$

Next, let us redo the problem, starting with $V(0) = 0$, and apply the improved Euler method with $\Delta t = 1.0$. As in the Euler method, we obtain a preliminary value at the end of the step of $V(1.0) = 8.80$, but now we shall evaluate the derivative there:

$$\frac{dV}{dt}\Big|_{1.0} = 8.80 - 1.37 \times 10^{-3} V^2 = 8.80 - 1.37 \times 10^{-3} * (8.80)^2 = 8.69.$$

Then,

$$V(1.0) = V(0) + \frac{1}{2}\left[\frac{dV}{dt}\Big|_t + \frac{dV}{dt}\Big|_{t+\Delta t}\right]\Delta t.$$

$$= 0 + \frac{1}{2}[8.80 + 8.69] * 1.0 = 8.75.$$

Again, the correct value here is 8.77, so we have a true relative error of 0.228%. This error, with the Euler method and the same step size, was 0.342%. This is an improvement, though we did almost twice as much work. Often more improvement will occur.

The next step begins with

$$\frac{dV}{dt}\Big|_{1.0} = 8.80 - 1.37 \times 10^{-3} \, V^2 = 8.80 - 1.37 \times 10^{-3} * (8.75)^2 = 8.70.$$

Notice that this is different from the derivative previously computed here, which was based on the single Euler step across the interval. At the end of the step, V is

$$V(2.0) = V(1.0) + \frac{dV}{dt}\Big|_{1.0} \Delta t = 8.75 + 8.70 * 1.0 = 17.5,$$

and the derivative there is

$$\frac{dV}{dt} = 8.80 - 1.37 \times 10^{-3} \, V^2 = 8.80 - 1.37 \times 10^{-3} * (17.5)^2 = 8.38.$$

Finally, using the derivatives at the end and beginning of the step, V is

$$V(2.0) = V(1.0) + \frac{1}{2}\left[\frac{dV}{dt}\Big|_{1.0} + \frac{dV}{dt}\Big|_{2.0}\right]\Delta t.$$

$$= 8.75 + 0.5\,[8.70 + 8.38] * 1.0 = 17.3.$$

Here, the correct answer is 17.3, so the true relative error is 0.0%, which is much better than the value of 1.16% obtained with the Euler method. Using the improved Euler method is beginning to pay off.

An alternate second-order method to the improved Euler method is the modified Euler method. (Note the similarity of the names and the potential confusion of the methods.) It begins like the Euler method and the improved Euler method but it only goes halfway across the interval instead of all the way. At the midpoint, the derivative is evaluated again. Then this derivative value is applied across the entire step. To continue our business forecast analogy, it is as if an estimate of the year based on June 30th earnings were taken as representative for the entire year. Here is the modified Euler method:

$$V(t + \frac{\Delta t}{2}) = V(t) + \frac{dV}{dt}\Big|_t \frac{\Delta t}{2};$$

then,

$$V(t + \Delta t) = V(t) + \frac{dV}{dt}\Big|_{t + \frac{\Delta t}{2}} \Delta t.$$

■ 8.5 Second-Order Equations

The equation we have solved so far has been first order in the velocity. What if we also want to know the position of the vehicle as a function of time, $x(t)$? From physics, we know that $\dfrac{dx}{dt} = V$, the velocity. Therefore, we have another first-order differential equation to solve, albeit a simple one. If we again employ Euler's method, we can write

$$x(t + \Delta t) = x(t) + \frac{dx}{dt} \Delta t$$

or, in this case,

$$x(t + \Delta t) = x(t) + V * \Delta t.$$

Suppose that at $t = 0$, $V = 0$ as before and, also, $x = 0$. Then

$$\frac{dx}{dt} = V(0) = 0$$

and

$$x(1.0) = x(0) + \frac{dx}{dt}(1.0) = 0 + 0 * (1.0) = 0.$$

We have the physically impossible result, assuming the brakes are off, that 1.0 seconds after the vehicle begins to accelerate, it is still at its initial position. (This sort of thing cannot happen except in cartoons when someone steps off a cliff and hovers midair for a few seconds before plunging downward.)

Although our arithmetic is correct, the problem is the result of applying a constant time derivative of 0, the initial velocity, across the entire step. Since the velocity immediately begins to increase for t greater than 0, using $V = 0$ across the step is physically incorrect. Using more evaluations of the derivative within the step, as we shall do shortly, will correct this problem.

The correct solution for x with $x(0) = 0$ is

$$x = 730.0 * \ln(\cosh(0.1098t)), \tag{8.5.1}$$

which is not an obvious result. At $t = 1.0$, this yields $x = 4.39$ ft. Hence, the true relative error at $t = 0.5$ is 100%. Not a good beginning!

Continuing the Euler solution at $t = 1.0$, we know $V(1.0) = 8.80$ from solving the velocity equation with the Euler method. Hence,

$$x(2.0) = x(1.0) + \frac{dx}{dt} * (1.0) = 0 + 8.80 * (1.0) = 8.80.$$

The correct value there, from Equation (8.5.1), is 17.5. So the true relative error at $t = 2.0$ is 49.7%, down a lot from the first step but not satisfying. We need to use a better method.

Before doing so, we should generalize the procedure using these next two equations. We have

$$\frac{dV}{dt} = 8.80 - 1.37 \times 10^{-3} \, V^2$$

and

$$\frac{dx}{dt} = V.$$

To solve the problem for x, we must solve both these equations. In this case, we can solve the V-derivative equation without using the x-derivative equation because only V appears on the right side of either equation. In general, such equations will involve both dependent variables on the right-hand side, so both equations have to be solved together. We shall do such an example shortly.

Since V is the time derivative of x, we could have written the V-derivative equation using a second derivative,

$$\frac{dV}{dt} = \frac{d\left(\frac{dx}{dt}\right)}{dt} = \frac{d^2x}{dt^2} = 8.80 - 1.37 \times 10^{-3} \left(\frac{dx}{dt}\right)^2$$

where we have removed V from the equation entirely by replacing it with $\frac{dx}{dt}$. Thus, we have a second-order equation in x.

Although this is a very common equation form, we cannot directly solve it numerically. We *must* break it up into two first-order equations, in this case, by replacing $\frac{dx}{dt}$ with V. Although this seems easy enough in principle, it tends to be confusing. If we are given any second-order equation, say

$$\frac{d^2p}{dq^2} = a * p + b * \frac{dp}{dq} + c * q + d$$

where p is the dependent variable and q is independent, introduce a new variable (pick your favorite letter) and let it stand for the first derivative of the dependent variable. In our example problem this was V, which had a familiar physical interpretation as the time derivative of x. If no such interpretation exists, do not worry about it—just think of it as an auxiliary variable. For example, let

$$w = \frac{dp}{dt}.$$

Then the second-order equation may be written as two first-order equations that are solved together.

$$\frac{dp}{dq} = w \quad \text{and} \quad \frac{d^2p}{dq^2} = \frac{d}{dq}\left(\frac{dp}{dq}\right) = \frac{dw}{dq} = a * p + b * w + c * q + d.$$

Do not attempt to integrate the second-order equation once to make it a first-order equation. In general, this is not mathematically possible. In our example,

$$\frac{d^2x}{dt^2} = 8.80 - 1.37 \times 10^{-3}\left(\frac{dx}{dt}\right)^2.$$

If we tried to integrate this once, the right-hand side would require that we integrate $\left(\frac{dx}{dt}\right)^2$ with respect to t. But we do not know the t-dependence of x (which is what we are trying to find numerically), so we cannot perform the integration. What we must do now is introduce the auxiliary variable to create another first-order equation and solve them both.

If we had a third-order equation we would have to make three first-order equations, and so on. We shall not pursue this but note that a famous equation for boundary layer flow over a flat plate is of this form.

$$ff'' + 2f''' = 0,$$

where f is related to the stream function and is a function of the distance from the plate.

■ 8.6 The Fourth-Order Runge–Kutta Method

We have seen how the Euler method and its relatives can be used to solve ordinary differential equations, but the results have not always been as good as we would like, especially for x. The Euler method is a first-order method; the improved Euler and the modified Euler methods are second-order methods—that is, they both agree with a Taylor series through the second derivative terms.

Let us now introduce a standard method for solving ordinary differential equations that is a fourth-order method; that is, it agrees with a Taylor series through the fourth derivative term. The standard notation for the various derivatives in this method is k with subscripts. The method actually uses a first derivative four times across the interval instead of the fourth derivative itself. The first evaluation of the derivative in the step, k_1, is evaluated at the beginning of the step. The second and third evaluations, k_2 and k_3, are evaluated at the middle of the step, and the last derivative evaluation, k_4, is at the end of the step. In a sense, the fourth-order Runge–Kutta method is a combination of the several Euler methods we have considered, since it evaluates the derivatives at the beginning, middle, and end of the interval.

The four derivatives are then weighted and combined to calculate the dependent variable at the end of the step. The weighting is heavier at the middle. The formula for the method is

$$V(t + \Delta t) = V(t) + \frac{\Delta t}{6} * [k_1 + 2k_2 + 2k_3 + k_4].$$

Since the two middle k's, k_2 and k_3, are both multiplied by 2, this weighting formula can be written a bit more compactly as

$$V(t + \Delta t) = V(t) + \frac{\Delta t}{6} * [k_1 + 2 * (k_2 + k_3) + k_4]. \qquad (8.6.1)$$

The method proceeds as follows:

1. Evaluate k_1 at the beginning of the step.
2. Use k_1 to advance the variable halfway across the step to calculate a temporary variable we shall call V_t. If more than one dependent variable is involved, their equations must be advanced together.
3. Use V_t to evaluate the derivative denoted k_2, at the *middle* of the interval—that is, at $t + \dfrac{\Delta t}{2}$.
4. Use k_2, applied at the *beginning* of the step to advance the dependent variable halfway across the step a second time. Call this estimate V_{tt}.
5. Use V_{tt} to calculate the derivative of V again and call the result k_3.
6. Use k_3, again applied at the beginning of the step, to advance the variable *all the way* across the step. Let the result at $t + \Delta t$ be called V_{ttt}.
7. Use V_{ttt} to calculate the derivative, called k_4, at the end of the step—that is, at $t + \Delta t$.
8. Combine these results using Equation (8.6.1).

The projections are shown in Figure 8.2.

We have perhaps belabored this process, but many texts write this procedure correctly although compactly—probably too compactly for someone unfamiliar with the method.

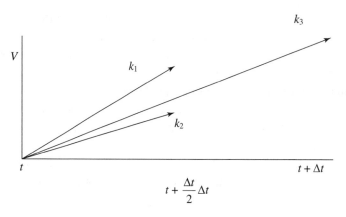

I FIGURE 8.2 Fourth-Order Runge–Kutta Derivative Projection.

8.6.1 The Example Again

We shall now solve the example problem for both x and V using the fourth-order Runge–Kutta method. To do this, we shall make two columns, with the dependent variable, V, on the left and the dependent variable, x, on the right. We shall use a double subscript on the k's to indicate the k value and the variable—for example, $k_{1,v}$. Again, $V(0) = 0$, $x(0) = 0$, and $\Delta t = 1.0$.

V side   **x side**

$$k_{1,V} = \frac{dV}{dt} = 8.80 - 1.37 \times 10^{-3}\, V^2 \qquad k_{1,x} = \frac{dx}{dt} = V$$

$$k_{1,V} = 8.80 - 1.37 \times 10^{-3}\, (0)^2 \qquad\qquad k_{1,x} = V(0) = 0$$

$$k_{1,V} = 8.80 \qquad\qquad\qquad\qquad\qquad\qquad k_{1,x} = 0$$

$$V(0.50) = V(0) + k_{1,V} * \frac{\Delta t}{2} \qquad\qquad x(0.50) = x(0) + k_{1,x} * \frac{\Delta t}{2}$$

$$V_t = 0 + 8.80 * 0.50 = 4.40 \qquad\qquad x_t = 0 + 0 * 0.50 = 0.$$

At this point we have advanced halfway across the first step. Continuing on:

V side   **x side**

$$k_{2,V} = 8.80 - 1.37 \times 10^{-3}\, (4.40)^2$$

$$k_{2,V} = 8.773 \qquad\qquad\qquad\qquad\qquad k_{2,x} = V_t = 4.40$$

$$V(0.5) = V(0) + k_{2,v} * \frac{\Delta t}{2} \qquad\qquad x(0.5) = x(0) + k_{2,x} * \frac{\Delta t}{2}$$

$$V_{tt} = 0 + 8.773 * 0.50 = 4.387 \qquad\qquad x_{tt} = 0 + 4.40 * 0.50 = 2.20$$

$$k_{3,V} = 8.80 - 1.37 \times 10^{-3}\, (4.387)^2$$

V side (cont.)

$k_{3,V} = 8.774$

$V(1.0) = V(0) + k_{3,V} * \Delta t$

$V_{ttt} = 0 + 8.774 * 1.0 = 8.774$

x side (cont.)

$k_{3,x} = V_{tt} = 4.387$

$x(1.0) = x(0) + k_{3,x} * \Delta t$

$x_{ttt} = 0 + 4.387 * 1.0 = 4.387$

We have now advanced all the way across the step. Continuing:

$k_{4,V} = 8.80 - 1.37 \times 10^{-3}(8.774)^2$

$k_{4,V} = 8.695$ $\qquad\qquad\qquad\qquad\qquad$ $k_{4,x} = V_{ttt} = 8.774$

Since all the k's are calculated we can apply Equation 8.6.1.

$$V(t + \Delta t) = V(t) + \frac{\Delta t}{6} * [k_{1,V} + 2 * (k_{2,V} + k_{3,V}) + k_{4,V}]$$

$$V(1.0) = V(0) + \frac{1.0}{6} * [8.80 + 2 * (8.773 + 8.774) + 8.695] = 8.765$$

$$x(t + \Delta t) = x(t) + \frac{\Delta t}{6} * [k_{1,x} + 2 * (k_{2,x} + k_{3,x}) + k_{4,x}]$$

$$x(1.0) = x(0) + \frac{1.0}{6} * [0 + 2 * (4.40 + 4.387) + 8.774] = 4.391.$$

The correct value of $V(1.0)$ is 8.765, so we have a correct solution to four significant figures. The correct value of $x(1.0)$ is 4.392, so we have a true error of 0.023%. This is much better than the results from the improved Euler method that we obtained earlier.

Now we shall do the second full step, beginning with

V side

$k_{1,V} = 8.80 - 1.37 \times 10^{-3}(8.765)^2$

$k_{1,V} = 8.695$

$V(1.50) = V(1.0) + k_{1,V} * \dfrac{\Delta t}{2}$

$V_t = 8.765 + 8.695 * 0.50 = 13.11$

$k_{2,V} = 8.80 - 1.37 \times 10^{-3}(13.11)^2$

$k_{2,V} = 8.565$

$V(1.50) = V(1.0) + k_{2,V} * \dfrac{\Delta t}{2}$

x side

$k_{1,x} = V(1.0) = 8.765$

$x(1.50) = x(1.0) + k_{1,x} * \dfrac{\Delta t}{2}$

$x_t = 4.391 + 8.765 * 0.50$

$x_t = 8.774$

$k_{2,x} = V_t = 13.11$

$x(1.50) = x(1.0) + k_{2,x} * \dfrac{\Delta t}{2}$

V side (cont.)

$V_{tt} = 8.765 + 8.565 * 0.50 = 13.05$

x side (cont.)

$x_{tt} = 4.391 + 13.11 * 0.50$

$x_{tt} = 10.95$

$k_{3,V} = 8.80 - 1.37 \times 10^{-3}(13.05)^2$

$k_{3,V} = 8.567$

$k_{3,x} = V_{tt} = 13.05$

$V(2.0) = V(1.0) + k_{3,V} * \Delta t$

$x(2.0) = x(1.0) + k_{3,x} * \Delta t$

$V_{ttt} = 8.765 + 8.567 * 1.0 = 17.33$

$x_{ttt} = 4.391 + 13.05 * 1.0$

$x_{ttt} = 17.44$

$k_{4,V} = 8.80 - 1.37 \times 10^{-3} * (17.33)^2$

$k_{4,V} = 8.389$

$k_{4,x} = V_{ttt} = 17.33$

with all the k's calculated we can now again apply Equation 8.6.1.

$$V(t + \Delta t) = V(t) + \frac{\Delta t}{6} * [k_{1,V} + 2 * (k_{2,V} + k_{3,V}) + k_{4,V}]$$

$$V(2.0) = 8.765 + \frac{1.0}{6} * [8.695 + 2 * (8.565 + 8.567) + 8.389] = 17.32$$

$$x(t + \Delta t) = x(t) + \frac{\Delta t}{6} * [k_{1,x} + 2 * (k_{2,x} + k_{3,x}) + k_{4,x}]$$

$$x(2.0) = 4.391 + \frac{1.0}{6} * [8.765 + 2 * (13.11 + 13.05) + 17.33] = 17.46.$$

The correct value of $V(2.0)$ is 17.32, so we again have the correct solution to four significant figures. The correct value of $x(2.0)$ is 17.46, so it is also correct.

But without knowing the correct values, how would we know if our answers are any good? And if we knew the correct answers, we certainly would not perform all these tedious calculations. Unfortunately, the fourth-order Runge–Kutta method does not provide an error estimate. Assuming no computational mistakes are made, the only source of error, aside from round-off, would come from an inappropriate step size. A rule-of-thumb for the step size in the fourth-order Runge–Kutta method is the Collatz criterion (see McCracken and Dorn, 1961), which addresses this question. This criterion requires that the following ratio involving the k's be a "few" hundredths or smaller. If it is not, the common step size should be reduced for all variables. The Collatz criterion is

$$\left| \frac{k_2 - k_3}{k_1 - k_2} \right|.$$

The criterion should be applied to all dependent variables used. If the step size is too big for any one variable, it should be reduced. The variables advance as a team.

Applying this criterion to the two steps of the example just worked, we find:

Step 1, V: $\left| \dfrac{k_2 - k_3}{k_1 - k_2} \right| = \left| \dfrac{8.773 - 8.774}{8.8 - 8.773} \right| = 0.0370,$

Step 1, x: $\left| \dfrac{k_2 - k_3}{k_1 - k_2} \right| = \left| \dfrac{4.4 - 4.387}{0 - 4.4} \right| = 0.00295,$

Step 2, V: $\left| \dfrac{k_2 - k_3}{k_1 - k_2} \right| = \left| \dfrac{8.565 - 8.567}{8.695 - 8.565} \right| = 0.0154,$

Step 2, x: $\left| \dfrac{k_2 - k_3}{k_1 - k_2} \right| = \left| \dfrac{13.11 - 13.05}{8.765 - 13.11} \right| = 0.0138.$

Thus the Collatz criterion is satisfied. Often the first step or two provide unreliable Collatz values, so a decision made immediately might be unrepresentative.

TIP

If this criterion is included in a computer program, a very small quantity, say 10^{-10}, should be added to the denominator. This will not affect results. However, if k_1 and k_2 are equal, as may happen at the beginning of a solution, division by zero would result without the inclusion of this small term and the solution would stop.

■ 8.7 Another Example with Coupled Equations

In the previous example, the equations were not mutually dependent on one another. Thus we were able to solve the V equation completely without considering the x equation. Now we shall consider a coupled set. We shall also change the names of the variables. This does not accomplish anything, but sometimes using variable names different from a familiar example causes confusion. This time x is the independent variable while w and z are the dependent variables. We shall arbitrarily take $w(0) = 1$, $z(0) = 2$, and $\Delta x = 0.1$.

The equations are as follows:

W side

$$k_{1,w} = \frac{dw}{dx} = -4 * w + 2 * z$$

$$k_{1,w} = 4 * 1 - 2 * 2 = 0$$

$$w(0.05) = w(0) + k_{1,w} * \frac{\Delta x}{2}$$

$$w_t = 1 + 0 * 0.05 = 1.000$$

Z side

$$k_{1,z} = \frac{dz}{dx} = 3 * w - 5 * z$$

$$k_{1,z} = 3 * 1 - 5 * 2 = -7$$

$$z(0.05) = z(0) + k_{1,z} * \frac{\Delta x}{2}$$

$$z_t = 2 - 7 * 0.05 = 1.650$$

W side (cont.)

$k_{2,w} = -4 * w_t + 2 * z_t$

$k_{2,w} = -4 * 1 + 2 * 1.65 = -0.7000$

$w(0.05) = w(0) + k_{2,w} * \dfrac{\Delta x}{2}$

$w_{tt} = 1 - 0.7 * 0.05 = 0.9650$

$k_{3,w} = -4 * w_{tt} + 2 * z_{tt}$

$k_{3,w} = -4 * 0.965 + 2 * 1.738$

$k_{3,w} = -0.3840$

$w(0.1) = w(0) + k_{3,w} \Delta x$

$w_{ttt} = 1 - 0.3840 * 0.1 = 0.9616$

$k_{4,w} = -4 * w_{ttt} + 2 * z_{ttt}$

$k_{4,w} = -4 * 0.9616 + 2 * 1.421$

$k_{4,w} = -1.004$

Z side (cont.)

$k_{2,z} = 3 * w_t - 5 * z_t$

$k_{2,z} = 3 * 1 - 5 * 1.65 = -5.250$

$z(0.05) = z(0) + k_{2,z} * \dfrac{\Delta x}{2}$

$z_{tt} = 2 - 5.25 * 0.05 = 1.738$

$k_{3,z} = 3 * w_{tt} - 5 * z_{tt}$

$k_{3,z} = 3 * 0.965 - 5 * 1.738$

$k_{3,z} = -5.795$

$z(0.1) = z(0) + k_{3,z} \Delta x$

$z_{ttt} = 2 - 5.795 * 0.1 = 1.421$

$k_{4,z} = 3 * w_{ttt} - 5 * z_{ttt}$

$k_{4,z} = 3 * 0.9616 - 5 * 1.421$

$k_{4,z} = -4.220$

$$w(x + \Delta x) = w(x) + \frac{\Delta x}{6} * [k_{1,w} + 2 * (k_{2,w} + k_{3,w}) + k_{4,w}]$$

$$w(0.1) = 1 + \frac{0.1}{6} * [0 + 2 * (-0.7 - 0.384) - 1.004] = 0.9471$$

$$z(x + \Delta x) = z(x) + \frac{\Delta x}{6} * [k_{1,z} + 2 * (k_{2,z} + k_{3,z}) + k_{4,z}]$$

$$z(0.1) = 2 + \frac{0.1}{6} * [-7 + 2 * (-5.25 - 5.795) - 4.22] = 1.445.$$

The correct value to four significant figures of $w(0.1)$ is 0.9476, so w from the fourth-order Runge–Kutta method differs in the fourth significant figure. The correct value of $z(0.1)$ is 1.444, so z achieves the same accuracy as w.

Let us apply the Collatz criterion to this problem:

For w, $\left| \dfrac{k_2 - k_3}{k_1 - k_2} \right| = \left| \dfrac{-0.7 - (-0.384)}{0 - (-0.7)} \right| = 0.45,$

and for z, $\left| \dfrac{k_2 - k_3}{k_1 - k_2} \right| = \left| \dfrac{-5.25 - (-5.795)}{-7 - (-0.5.25)} \right| = 0.31.$

These results are more than just a few hundredths, yet the solution was accurate. This demonstrates that the Collatz criterion is conservative because it is difficult to satisfy.

The results are good although the Collatz values are not. A smaller stepsize should have been used according to the criterion.

■ 8.8 Two-Point Boundary Value Problems

Sometimes in solving two or more simultaneous differential equations we have the needed number of initial conditions, but they are not all at the same location. For example, in solving a counterflow heat exchanger problem, we might know the inlet temperatures of the two fluids but they are at opposite ends of the exchanger. Our solution scheme must march in one direction or the other—that is, we must start at $x = 0$ or $x = L$, the length. Wherever we start, we need temperatures for both dependent variables. In another problem, we might know the initial position but the specified velocity is at some subsequent location. Such problems are referred to as two-point boundary value problems. A general schematic is shown in Figure 8.3.

These differences in location pose no problems for closed-form solutions because we can evaluate the constants wherever we have given information. But for numerical solutions, we must guess the missing condition(s) at the starting point we select, and then we advance the solution toward the other end. When we advance to the location where the other initial condition is specified, we can compare the computed and specified conditions. Then we can adjust our guessed starting value(s) and repeat—iterate—the process. Perhaps surprisingly, the process just described works quite well. Because we are trying to hit a target with this technique, this process is called *shooting*.

As an example, let us consider again our accelerating vehicle. To shorten the example considerably, we assume the position is known at the beginning of the first step but the velocity is known at the end of the first step. In reality, the two conditions would be much farther apart than this. Solving this small problem will illustrate the technique without compromising its generality and require many fewer time steps.

Suppose $x(0) = 0$ and $V(1.0) = 10.0$. Previously, we found that when $V(0) = 0, V(1.0) = 8.765$. Hence, we would reasonably assume that $V(0)$ must be greater than 0 for the new condition, and so we will assume arbitrarily that $V(0) = 1.0$. When we solve the problem with this $V(0)$, we obtain the following k's and velocity at $t = 1.0$:

$$k_1 = 8.800 \quad k_2 = 8.773 \quad k_3 = 8.774 \quad k_4 = 8.695 \quad V(1.0) = 9.751.$$

$x = 0$ _____ $x = L$
$\quad\quad$ $T_1(0)$ known $\quad\quad\quad\quad\quad\quad\quad\quad\quad\quad\quad\quad$ $T_2(L)$ known
$\quad\quad\quad\quad\quad\quad$ solution march =>

I FIGURE 8.3 Schematic of Two-Point Boundary Value Problem for Variables T_1 and T_2.

Thus our guess was in the right direction, but it fell short. We shall have to try another value for $V(0)$ greater than 1. We might randomly try several values until we hit what we wanted, or we might try to ratio our error and guess: that is,

$$\frac{\text{target}}{\text{firstresult}} * \text{firstguess} = \frac{10}{9.751} * 1 = 1.026$$

and use this for $V(0)$.

But there is a better way. We can let the history of our results predict the next guess by constructing a one-derivative Taylor series (them again). This will be possible once we have two consecutive results so that we may approximate the derivative of the unknown function numerically. This is not a serious restriction—we can make an arbitrary first guess and then increase it or decrease it depending on whether our answer falls short of, or exceeds, the desired value. In our current problem, our first guess of V at $t = 0$ was low, so we should increase the value of the second guess.

Let V_{target} be the value we are trying to match, let V^i be the result of the ith iteration, and G^i be the ith guess. Then

$$V_{\text{target}} \approx V^i + \frac{V^i - V^{i-1}}{G^i - G^{i-1}} * (G^{i+1} - G^i).$$

Only G^{i+1} will be unknown after the ith iteration so we can solve for it. V_{target} is fixed. Presently we have $V_{\text{target}} = 10.0, G^1 = 0, V^1 = 8.765, G^2 = 1$, and $V^2 = 9.751$. So

$$10.0 = 9.751 + \frac{9.751 - 8.765}{1 - 0} * (G^3 - 1).$$

Solving this equation, we calculate $G^3 = 1.253$, and when this is used in the fourth-order Runge–Kutta method, we find $V(1.0) = 10.00$. This is our target value. This was unusually rapid convergence, the result of our two values being separated by only one full step. Usually, several more iterations would be expected before we obtain a result sufficiently close to our target to quit. The process is quite automatic and relatively easy to program, though we have to be careful when indexing our new and old values.

If shooting does not work well, we can use a technique called *relaxation*. This is different from the relaxation we shall introduce in trying to speed up the rate of convergence when applying Gauss–Seidel to sets of linear equations in Chapter 9. Unfortunately, in numerical analysis, relaxation has a number of quite different meanings. In the context of boundary value problems, relaxation requires that we construct a set of finite difference equations and then solve them by advancing from the known to the unknown points. However, the resulting equations are often nonlinear; therefore, numerical techniques like the secant rule might have to be employed at each point to solve them. The relaxation technique might work when shooting does not, but it will probably be more complicated, so the shooting method is preferred. Press et al. (1989) made the memorable recommendation in their book that "As old computer gunslingers, we say shoot first and then relax."

■ 8.9 A Predictor–Corrector Method

Another class of methods for solving ordinary differential equations is called predictor–corrector methods. These methods iterate on the solution over a given step (even when everything is known at the beginning of the step). Comparing the changes that result during the iteration process provides an estimate of the error of the solution. This is a distinct advantage. On the other hand, these methods have several disadvantages compared to the Euler and Runge–Kutta methods. We introduce the method to familiarize the student with the concept. In Section 8.10, we cite a method that combines the general ease of use of the Euler and Runge–Kutta methods with an error estimate.

Recall the improved Euler method, a second-order method. We first calculated the derivative at the beginning of the step, used this to project across the step, and then computed the derivative at the end of the step. Then we averaged the two derivatives and used the average derivative to compute the value at the end of the step.

There is nothing to prevent us from using this value to compute a second estimate of the derivative at the end of the step and then using this in our average. This would be a simple predictor–corrector method. However, the first time the derivative at the end of the step is calculated as described, it is based on an Euler projection, which is only first-order accurate. This, unfortunately, restricts the accuracy of subsequent calculations. What we would like to have is a second-order estimate to evaluate our derivative at the end of the step. This estimate is called the predictor, and in the hierarchy of results it is denoted with superscript zero: for example, V^0. This simple second-order predictor makes use of the results from two consecutive steps:

$$V^0(t + \Delta t) = V(t - \Delta t) + 2 * \left.\frac{dV}{dt}\right|_t * \Delta t.$$

Notice that we proceed from the value before the present time: that is, $V(t - \Delta t)$ by using the derivative at t and applying it across two full steps. This gives us a second-order accurate predictor but it comes at a price. We need two consecutive values in order to use it and hence the method is not self-starting. We need to use another method, such as improved Euler, or fourth-order Runge–Kutta, and so on, to get started. Furthermore, if we should decide to change the step size, we would have to back up to the last known value and reintroduce our starting method in order to again calculate two consecutive known values equally separated by the new step size. Thus we begin to see the complications of using these methods.

Once we have our predicted value, $V^0(t + \Delta t)$, we can use it to get the derivative at $t + \Delta t$, $\left. \dfrac{dV}{dt} \right|_{t+\Delta t}$. Then the first corrected value at $t + \Delta t$, V^1 may be calculated as follows:

$$V^1 = V(t) + 0.5 * \left[\left. \frac{dV}{dt} \right|_t + \left. \frac{dV}{dt} \right|_{t+\Delta t} \right] * \Delta t. \tag{8.9.1}$$

This, in turn, may be used to calculate another value for the derivative at the end of the step and the result used in Equation (8.9.1) to generate V^2. In general,

$$V^{i+1}_{t+\Delta t} = V(t) + 0.5 * \left[\left. \frac{dV}{dt} \right|_t + \left. \frac{dV^i}{dt} \right|_{t+\Delta t} \right] * \Delta t$$

where we have used a superscript on the V in the derivative to show which value of V is used to compute the derivative. Continuation should lead to a converged value though not necessarily to the correct value. But, if more than two corrections are needed to satisfy an error criterion, it is probably more efficient to reduce the step size than to keep refining the result with the given step size.

The error estimate is made by comparing the current corrected value, V^i, with the predicted value, V^0. For the present method the error is

$$E_c = -0.2 * [V^i - V^0].$$

There are more accurate (and complicated) predictor–corrector methods that involve more points in the process, as for example, the fourth-order Adams–Moulton method that uses three previous points to calculate the value at the new point (see Mc-Cracken and Dorn, 1961).

We shall apply our method to the vehicle acceleration problem with $\Delta t = 0.5$. Since the predictor–corrector method is not self-starting, we shall use the results from the first step of the fourth-order Runge–Kutta method at $t = 1.0$, $V(1.0) = 8.765$, and $x(1.0) = 4.391$.

V side

$$\frac{dV}{dt} = 8.80 - 1.37 \times 10^{-3}\, V^2$$

$$\left.\frac{dV}{dt}\right|_{t=1.0} = 8.80 - 1.37 \times 10^{-3}(8.765)^2$$

$$\left.\frac{dV}{dt}\right|_{t=1.0} = 8.695$$

$$V^0(2.0) = V(0) + 2 * \left.\frac{dV}{dt}\right|_{1.0} * \Delta t$$

$$V^0(2.0) = 0 + 2 * 8.695 * 1.0$$
$$V^0(2.0) = 17.39$$

$$\left.\frac{dV}{dt}\right|_{2.0} = 8.80 - 1.37 \times 10^{-3}(17.39)^2$$

$$\left.\frac{dV}{dt}\right|_{2.0} = 8.386$$

$$V^1(2.0) = V(1.0) + 0.5 * \left[\left.\frac{dV}{dt}\right|_{1.0} + \left.\frac{dV^0}{dt}\right|_{2.0}\right] * \Delta t$$

$$V^1(1.0) = 8.765 + 0.5 * (8.695 + 8.386) * 1.0 = 17.31$$

$$\left.\frac{dV^1}{dt}\right|_{2.0} = 8.80 - 1.37 \times 10^{-3}(17.31)^2$$

$$\left.\frac{dV^1}{dt}\right|_{2.0} = 8.390$$

$$V^2(2.0) = V(1.0) + 0.5 * \left[\left.\frac{dV}{dt}\right|_{1.0} + \left.\frac{dV^1}{dt}\right|_{2.0}\right] * \Delta t$$

$$V^2(2.0) = 8.765 + 0.5 * (8.695 + 8.390) * 1.0 = 17.31$$

x side

$$\frac{dx}{dt} = V$$

$$\left.\frac{dx}{dt}\right|_{t=1.0} = 8.765$$

$$x^0(2.0) = x(0) + 2 * \left.\frac{dx}{dt}\right|_{1.0} * \Delta t$$

$$x^0(2.0) = 0 + 2 * 8.765 * 1.0$$
$$x^0(2.0) = 17.53$$

$$\left.\frac{dx}{dt}\right|_{2.0} = 17.39$$

$$x^1(2.0) = x(1.0) + 0.5 * \left[\left.\frac{dx}{dt}\right|_{1.0} + \left.\frac{dx^0}{dt}\right|_{2.0}\right] * \Delta t$$

$$x^1(2.0) = 4.391 + 0.5 * (8.765 + 17.39) * 1.0 = 17.47$$

$$\left.\frac{dx^1}{dt}\right|_{2.0} = 17.31$$

$$x^2(2.0) = x(1.0) + 0.5 * \left[\left.\frac{dx}{dt_0}\right|_{1.0} + \left.\frac{dx^1}{dt}\right|_{2.0}\right] * \Delta t$$

$$x^2(2.0) = 4.391 + 0.5 * (8.765 + 17.31) * 1.0 = 17.43.$$

Note that the derivatives at the beginning of the step do not change.

Since the correct values for $V(2.0)$ and $x(2.0)$ are 17.32 ft/s and 17.46 ft, respectively, the true error for V is 0.03 ft/s and for x, 0.03 ft. Applying the error estimate formula for this method, we obtain for V:

$$E_{c, V} = -0.2 * [V^2 - V^0] = -0.2 * (17.31 - 17.39) = 0.016 \text{ ft/s},$$

and for x,

$$E_{c, x} = -0.2 * [x^2 - x^0] = -0.2 * (17.43 - 17.53) = 0.02 \text{ ft}.$$

The error estimates are close, but both are lower than the true errors. Nevertheless, they provide us with an estimate of the errors.

■ 8.10 The Cash–Karp Runge–Kutta Method

A method similar to the fourth-order Runge–Kutta method that provides both an error estimate and is self-starting is the Cash–Karp Runge–Kutta method (see Cash and Karp, 1990). This is a reformulation of a method developed first by E. Fehlberg. To compute the new variable and the error requires the evaluation of derivatives (k's) at six irregular locations across the step (beginning, $0.2\Delta t$, $0.3\Delta t$, $0.6\Delta t$, $1.0\Delta t$, and $0.875\Delta t$) and their weighted combination. Because the weighting factors are complicated fractions, the method is not convenient for calculation by hand. But once programmed, it provides a very attractive method for solving ordinary differential equations. If the error is estimated periodically and compared to a tolerance, the step size can be adjusted as the method proceeds. Accordingly, we have the basis of a self-adaptive method. In regions of rapid change, the step size selected will be small; in regions of slow change, larger steps may be used safely.

The compact forms of the basic equations follow. Note they have been written for one dependent variable, V, and one independent variable, t. If a second dependent variable were in the problem, another argument would be added. The fourth-order result is in Equation (8.10.1); the fifth-order result is in (8.10.2). The error is estimated by subtracting the fourth-order result from the fifth-order result: that is, (8.10.2) minus (8.10.1). Notice that k_2 does not appear explicitly in either Equations (8.10.1) or (8.10.2), but it is needed to get the higher k's. Accordingly:

$$V(t + \Delta t) = V(t) + (\frac{37}{378} * k_1 + \frac{250}{621} * k_3 + \frac{125}{594} * k_4 \qquad (8.10.1)$$

$$+ \frac{512}{1771} * k_6) * \Delta t \qquad \text{(fourth order)}$$

$$k_1 = f(t, V),$$
$$k_2 = f(t + 0.2 * \Delta t, V + 0.2 * k_1 * \Delta t),$$

$$k_3 = f(t + 0.3 * \Delta t, V + (\frac{3}{40} * k_1 + \frac{9}{40} * k_2) * \Delta t),$$

$$k_4 = f(t + 0.6 * \Delta t, V + (\frac{3}{10} * k_1 - \frac{9}{10} * k_2 + \frac{6}{5} * k_3) * \Delta t),$$

$$k_5 = f(t + \Delta t, V + (-\frac{11}{54} * k_1 + \frac{5}{2} * k_2 - \frac{70}{27} * k_3 + \frac{35}{27} * k_4) * \Delta t),$$

$$k_6 = f(t + 0.5 * \Delta t, V + (\frac{1631}{55,296} * k_1 + \frac{175}{512} * k_2 + \frac{575}{13,824} * k_3$$

$$+ \frac{44,275}{110,592} * k_4 + \frac{253}{4096} * k_5) * \Delta t).$$

The fifth-order result is given by

$$V(t + \Delta t) = V(t) + (\frac{2825}{27,648} * k_1 + \frac{18,575}{48,384} * k_3 + \frac{13,525}{55,296} * k_4$$

$$+ \frac{277}{14,336} * k_5 + \frac{1}{4} * k_6) * \Delta t. \tag{8.10.2}$$

The error is: $E_A = (8.10.2) - (8.10.1)$.

This is an absolute rather than a relative error. Periodic checking of E_A can lead to adjusting the step size in order to solve the equation as fast as possible while maintaining error control. This leads to an adaptive method and seems the best way to proceed.

Let us apply this procedure to the velocity example problem, where

$$\frac{dV}{dt} = 8.80 - 1.37 \times 10^{-3}V^2 \text{ with } V(0) = 0.$$

$$k_{1,V} = \frac{dV}{dt} = 8.80 - 1.37 \times 10^{-3} * 0^2 = 8.800,$$

$$V_t = V(0) + 0.2 * k_1 * \Delta t = 0 + 6.2 * 8.800 * 1 = 1.760,$$

$$k_{2,V} = 8.80 - 1.37 \times 10^{-3} * (1.760)^2 = 8.796,$$

$$V_{tt} = V(0) + (\frac{3}{40} * k_1 + \frac{9}{40} * k_2) * \Delta t = 0 + (\frac{3}{40} * 8.800 + \frac{9}{40} * 8.796)$$

$$* 1 = 2.639,$$

$$k_{3,V} = 8.80 - 1.37 \times 10^{-3} * (2.639)^2 = 8.790,$$

$$V_{ttt} = V(0) + (\frac{3}{10} * k_1 - \frac{9}{10} * k_2 + \frac{6}{5} * k_3) * \Delta t,$$

$$V_{ttt} = 0 + (\frac{3}{10} * 8.800 - \frac{9}{10} * 8.796 + \frac{6}{5} * 8.790) * 1 = 5.272,$$

$$k_{4,V} = 8.80 - 1.37 \times 10^{-3} * (5.272)^2 = 8.762,$$

$$V_{tttt} = V + (-\frac{11}{54} * k_1 + \frac{5}{2} * k_2 - \frac{70}{27} * k_3 + \frac{35}{27} * k_4) * \Delta t,$$

$$= 0 + (-\frac{11}{54} * 8.800 + \frac{5}{2} * 8.796 - \frac{70}{27} * 8.790 + \frac{35}{27} * 8.762) * 1 = 8.767,$$

$$k_{5,V} = 8.80 - 1.37 \times 10^{-3} * (8.767)^2 = 8.695,$$

$$V_{ttttt} = V + (\frac{1631}{55,296} * k_1 + \frac{175}{512} * k_2 + \frac{575}{13,824} * k_3 + \frac{44,275}{110,592} * k_4$$

$$+ \frac{253}{4096} * k_5) * \Delta t,$$

$$V_{ttttt} = 0 + (\frac{1631}{55,296} * 8.800 + \frac{175}{512} * 8.796 + \frac{575}{13,824} * 8.790$$

$$+ \frac{44,275}{110,592} * 8.762 + \frac{253}{4096} * 8.695) * 1 = 7.677,$$

$$k_{6,V} = 8.80 - 1.37 \times 10^{-3} * (7.677)^2 = 8.719.$$

Then the fourth-order result is given by Equation (8.10.1):

$$V(t + \Delta t) = V(t) + (\frac{37}{378} * k_1 + \frac{250}{621} * k_3 + \frac{125}{594} * k_4 + \frac{512}{1771} * k_6) * \Delta t,$$

$$V(1) = 0 + (\frac{37}{378} * 8.800 + \frac{250}{621} * 8.790 + \frac{125}{594} * 8.762$$

$$+ \frac{512}{1771} * 8.695) * 1 = 8.765.$$

The fifth-order result is given by Equation (8.10.2):

$$V(t + \Delta t) = V(t) + (\frac{2825}{27,648} * k_1 + \frac{18,575}{48,384} * k_3 + \frac{13,525}{55,296} * k_4$$

$$+ \frac{277}{14,336} * k_5 + \frac{1}{4} * k_6) * \Delta t,$$

$$V(1) = 0 + (\frac{2825}{27,648} * 8.800 + \frac{18,575}{48,384} * 8.790 + \frac{13,525}{55,296} * 8.762$$

$$+ \frac{277}{14,336} * 8.695 + \frac{1}{4} * 8.719) * 1 = 8.765.$$

This is the correct answer to four significant figures. The error estimate, $E_A = V(\text{fifth-order}) - V(\text{fourth-order}) = 8.765 - 8.765 = 0.000$, which is consistent.

Since $\dfrac{dx}{dt} = V$, the previous results give us $k_{1x} = 0$, $k_{2x} = V_t = 1.760$,

$k_{3x} = V_{tt} = 2.639$, $k_{4x} = V_{ttt} = 5.272$, $k_{5x} = V_{tttt} = 8.767$, and $k_{6x} = V_{ttttt} = 7.677$.

Then the fourth-order result is given by Equation (8.10.1):

$$x(t + \Delta t) = x(t) + (\frac{37}{378} * k_1 + \frac{250}{621} * k_3 + \frac{125}{594} * k_4 + \frac{512}{1{,}771} * k_6) * \Delta t,$$

$$x(1) = 0 + (\frac{37}{378} * 0.000 + \frac{250}{621} * 2.639 + \frac{125}{594} * 5.272$$

$$+ \frac{512}{1771} * 7.677) * 1 = 4.391.$$

The fifth-order result is given by Equation (8.10.2):

$$x(t + \Delta t) = x(t) + (\frac{2825}{27{,}648} * k_1 + \frac{18{,}575}{48{,}384} * k_3 + \frac{13{,}525}{55{,}296} * k_4$$

$$+ \frac{277}{14{,}336} * k_5 + \frac{1}{4} * k_6) * \Delta t,$$

$$x(1) = 0 + (\frac{2825}{27{,}648} * 0.000 + \frac{18{,}575}{48{,}384} * 2.639 + \frac{13{,}525}{55{,}296} * 5.272$$

$$+ \frac{277}{14{,}336} * 8.767 + \frac{1}{4} * 7.677) * 1 = 4.391.$$

The correct value is 4.392, so the true error is 0.001. Taking the difference between the fifth- and fourth-order results to four significant figures yields an error estimate of 0. Carrying more significant figures would have yielded a better error estimate.

8.10.1 VBA Program Results

A VBA program to utilize the Cash–Karp Runge–Kutta method for two ordinary differential equations is found in Appendix H. The computational step size is variable, depending on how the error terms compare to a tolerance. The differential equations are defined in functions f and g.

The results from applying this program to the velocity example problem for $0 <= t < 50$ are in Table 8.1. At each printout, the calculated errors in V and x were compared to a tolerance value of 1×10^{-5} (1E-5 in VBA). If both were less than one-tenth of this, the step size was doubled. If both were more than 10 times this, the step size was halved. Because of the adjustable step size, the last step took the solution from $t = 49$ to $t = 57$ secs with $\Delta t = 8.0$ secs. Comparing the numerical and correct solutions, we see that even with the large step size of 8.0 secs, the difference between the

Table 8.1 Cash–Karp and correct values with variable step size

t	V	x	Vcorr	xcorr	tfinal 50 eaV	pf 1 eax	dt
0	0	0	0	0	50	1	1
1	8.764806	4.391188	8.765273	4.391641	6.71E-08	–2.5E-07	1
3	25.48488	38.90389	25.48622	38.90795	9.21E-07	–3.4E-06	2
5	40.05451	104.8812	40.05659	104.8922	8.93E-07	–2.1E-06	2
7	51.78343	197.2124	51.78609	197.233	9.17E-07	–1.3E-06	2
9	60.6379	310.0887	60.64099	310.1211	8.64E-07	–8.7E-07	2
13	71.42231	576.796	71.42427	576.8636	2.59E-05	–1.5E-05	4
17	76.40378	873.7753	76.40578	873.8739	1.63E-05	–6.6E-06	4
21	78.57032	1184.326	78.57294	1184.454	8.14E-06	–2.5E-06	4
25	79.4878	1500.702	79.49103	1500.86	3.66E-06	–8.9E-07	4
29	79.87193	1819.531	79.87557	1819.719	1.57E-06	–3.2E-07	4
33	80.03199	2139.385	80.03588	2139.604	6.63E-07	–1.1E-07	4
41	80.12669	2780.12	80.13029	2780.406	1.24E-05	–1.6E-06	8
49	80.14263	3421.215	80.1466	3421.564	2.09E-06	–2.2E-07	8
57	80.14531	4062.369	80.14941	4062.783	3.51E-07	–3.2E-08	8

computed and correct solution appears in the fourth significant figure, which is the number retained in the correct solution. The correct solutions were calculated in the spreadsheet using equations (8.2.2) and (8.5.1).

■ 8.11 Stability

Errors caused by round-off and truncation are inevitable in numerical calculations. In solving ordinary differential equations, we endeavor to control errors by using appropriate step sizes and higher-order methods. But what about the errors that do creep in? Do they fade away or do they accumulate, like a snowball rolling downhill picks up snow as it rolls? If errors accumulate, they will eventually lead to meaningless results that destroy the solution. This is called instability, and that is the bad news. The good news is that unstable solutions lead to obviously bad results so we are not likely to use them without noticing. This idea was well-expressed by my former student who described instability as "Problems that produce nonsensical results when solved."

The methods we have used have always been of the following form, in which the derivative was a forward derivative—that is, evaluated where V was known. Thus we used a forward difference. Consequently,

$$\frac{dV}{dt}\Big|_t = \frac{V(t + \Delta t) - V(t)}{\Delta t},$$

which we proceeded to advance, leading to

$$V(t + \Delta t) = V(t) + \frac{dV}{dt}\Big|_t \Delta t.$$

This is an explicit form of the equation.

But suppose we had written our derivative as a backward derivative—that is, evaluated it at $t + \Delta t$:

$$\frac{dV}{dt}\Big|_{t+\Delta t} = \frac{V(t + \Delta t) - V(t)}{\Delta t}.$$

Then

$$V(t + \Delta t) = V(t) + \frac{dV}{dt}\Big|_{t+\Delta t} \Delta t,$$

and

$$V(t + \Delta t) - \frac{dV}{dt}\Big|_{t+\Delta t} \Delta t = V(t).$$

Now both $V(t + \Delta t)$ and $\frac{dV}{dt}\Big|_{t+\Delta t}$ are on the left side of the equation and hence unknown. This formulation is called *implicit*. It has stability advantages but is harder to use, often requiring the solution of nonlinear algebraic equations at each step. We shall not consider implicit methods again. They have frequent application in solving parabolic partial differential equations.

For explicit methods, a maximum step size (a solution "speed limit") exists for each method considered. Writing and solving a difference equation for the generic error leads to an expression for this limit. For the fourth-order Runge–Kutta method, a stable solution has two requirements for each dependent variable (see McCracken and Dorn, 1961). Label the derivative, $\frac{dV}{dt}$, as f. Then,

$$\frac{\delta f}{\delta V} \text{ must be negative, and } \frac{2.8}{\left|\dfrac{\delta f}{\delta V}\right|} > \Delta t.$$

The first of these two criteria means the derivative expression must depend negatively on the variable: that is, an increase in the variable must produce an algebraically negative change in the derivative. Thus, positive changes in the variable do not produce positive changes in the derivative and hence the possibility of unlimited growth. It is a measure of the sensitivity of the derivative to the dependent variable.

Recall our two derivatives for the vehicle acceleration problem:

$$\frac{dV}{dt} = 8.80 - 1.37 \times 10^{-3} V^2 \text{ and } \frac{dx}{dt} = V.$$

Then for V,

$$\frac{\delta f}{\delta V} = \frac{\delta(8.80 - 1.37 \times 10^{-3} * V^2)}{\delta V} = -2.74 \times 10^{-3} V,$$

which is always negative for positive V. Hence, the first criterion is satisfied. The second criterion depends on the value of V:

$$\frac{2.8}{|2.74 \times 10^{-3} * V|}.$$

Earlier, we obtained $V(1.0) = 8.765$ and $V(2.0) = 17.32$. Since the larger V yields the smaller fraction, evaluation at $V = 17.32$ yields

$$59.4 > \Delta t.$$

Clearly, this is not a problem since we were using a step size of only 1.0 seconds. But if we were to exceed the limit, the solution would begin to oscillate wildly. Solution failure would be obvious. Note that being stable does not mean the solution is accurate. Stability limits are weaker constraints than accuracy limits.

For the x variable, we have $f = \frac{dx}{dt} = V$. Then, taking the derivative of f with respect to the variable of interest, x, we find

$$\frac{\delta f}{\delta x} = \frac{\delta V}{\delta x} = 0.$$

There is no stability limit imposed by the x derivative equation because it does not contain x. It is not self-sensitive!

■ 8.12 Stiff Differential Equations

In Section 8.7, we easily obtained an accurate solution for the following pair of equations over one sample step:

$$\frac{dw}{dx} = -4 * w + 2 * z \qquad \frac{dz}{dx} = 3 * w - 5 * z,$$

$w(0) = 1, z(0) = 2$, and we used a step size of 0.1.

If the z derivative equation were altered slightly to become

$$\frac{dz}{dx} = 3 * w - 500 * z,$$

the new z coefficient of 500 would require that we use a much smaller step size to solve for z than for w. Sets where one or more variables hold back the progress of the others are called *stiff*. Stiff equations slow down the common step size Δt that must be applied to the whole set, just as someone who has trouble keeping up with a group can slow down the progress of the whole group.

The explanation for this behavior can be seen by considering the analytic solution to these equations. For the equation set of Section 8.7, these are

$$w = -0.4 * e^{(-7 * x)} + 1.4 * e^{(-2 * x)} \quad \text{and} \quad z = 0.6 * e^{(-7 * x)} + 1.4 * e^{(-2 * x)}.$$

(The common coefficient of 1.4 for the second term for each variable is a coincidence.)
The analytic solution of the new set is

$$w = -1.008 * e^{(-3.9879 * x)} - 0.0080 * e^{(-500.01 * x)} \text{ and}$$
$$z = 0.006088 * e^{(-3.9879 * x)} + 1.9939 * e^{(-500.01 * x)}.$$

The much smaller step size needed to solve this equation set is the result of the large negative coefficient in the second exponential.

Gear (see Gear, 1968) developed an implicit scheme for solving such equations. An alternate approach that is simpler to apply though less general is due to Hiestand (see Hiestand and George, 1976). Based on the steady state approximation of chemical kinetics, the approach proceeds as follows.

In the second equation of the altered set, the -500 as the coefficient of z means the value of z changes much more quickly than the value of w. Hence, if we take a step size suitable for w, z has probably reached a local equilibrium value: that is, its derivative has become zero. Instead of solving both equations, we solve the w-derivative conventionally in the equation for w. But instead of solving the z-derivative equation, we set $\frac{dz}{dx} = 0$.

Thus the second equation becomes $\frac{dz}{dx} = 0 = 3 * w - 500 * z$, which may be solved for z as

$$z = \frac{3 * w}{500}.$$

The characteristic time (or space), τ may be estimated as:

$$\tau_i = \frac{1}{\left| \dfrac{\delta}{\delta y_i} \left(\dfrac{\delta y_i}{\delta t} \right) \right|}.$$

In this example, $\tau_w = \dfrac{1}{4} = 0.25$ and $\tau_z = \dfrac{1}{500} = 0.002$, which is 125 times smaller than τ_w.

When this scheme was applied to 25 steps of this set with a step size of 0.01, the results shown in Table 8.2 were obtained. The table contains the computed results, the correct results, the true relative errors, and the Collatz criterion for w. Also shown in the right two columns of the table are the results of the calculation if a normal solution is attempted, without dealing with the stiffness. The results are clearly absurd.

The absurd results in the right two columns should not be surprising. The stability limit for the z equation follows.

$$\text{Let } f = \frac{dz}{dx} = 3 * w - 500 * z.$$

$$\text{Then } \frac{\delta f}{\delta z} = -500, \text{ and}$$

$$\frac{2.8}{\left| \dfrac{\delta f}{\delta V} \right|} = \frac{2.8}{500} = 0.0056 > \Delta x.$$

Hence our step size was twice what stability would allow, without even considering its effect on accuracy.

■ 8.13 Ordinary Differential Equation Methods and Numerical Integration

It is worth commenting on the general difference between numerical integration and the numerical solution of ordinary differential equations. We often casually say we are integrating a differential equation. Though this is possible in simple cases, in general it is not, and the two processes must not be confused.

Consider the differential equation,

$$\frac{dV}{dt} = f(V, t).$$

Table 8.2   Steady-state, correct, and stiff equation results

	with steady state							with stiffness	
x	w	z	wexact	zexact	Trerrw, %	Trerrz, %	Collatz	w	z
0	1	2							
0.01	0.973939	0.005844	0.968539	0.019283	−0.55753	69.69611	0.01994	0.858408	27.34192
0.02	0.935863	0.005615	0.930726	0.005712	−0.55196	1.69154	0.01994	−0.58038	374.7783
0.03	0.899276	0.005396	0.89434	0.005402	−0.5519	0.120032	0.01994	−19.8231	5138.073
0.04	0.86412	0.005185	0.859377	0.00519	−0.55188	0.108854	0.01994	−283.174	70442.01
0.05	0.830337	0.004982	0.82578	0.004987	−0.55186	0.108794	0.01994	−3893.22	965747.7
0.06	0.797876	0.004787	0.793497	0.004792	−0.55184	0.108812	0.01994	−53385.9	13240232
0.07	0.766683	0.0046	0.762476	0.004605	−0.55182	0.108831	0.01994	−731922	1.82E+08
0.08	0.73671	0.00442	0.732667	0.004425	−0.55181	0.108849	0.01994	−1E+07	2.49E+09
0.09	0.707909	0.004247	0.704024	0.004252	−0.55179	0.108868	0.01994	−1.4E+08	3.41E+10
0.1	0.680233	0.004081	0.676501	0.004086	−0.55177	0.108886	0.01994	−1.9E+09	4.68E+11
0.11	0.65364	0.003922	0.650053	0.003926	−0.55175	0.108905	0.01994	−2.6E+10	6.41E+12
0.12	0.628086	0.003769	0.62464	0.003773	−0.55173	0.108923	0.01994	−3.5E+11	8.79E+13
0.13	0.603532	0.003621	0.60022	0.003625	−0.55171	0.108942	0.01994	−4.9E+12	1.21E+15
0.14	0.579937	0.00348	0.576755	0.003483	−0.55169	0.108961	0.01994	−6.7E+13	1.65E+16
0.15	0.557265	0.003344	0.554207	0.003347	−0.55168	0.108979	0.01994	−9.1E+14	2.27E+17
0.16	0.535479	0.003213	0.532541	0.003216	−0.55166	0.108998	0.01994	−1.3E+16	3.11E+18
0.17	0.514544	0.003087	0.511721	0.003091	−0.55164	0.109016	0.01994	−1.7E+17	4.26E+19
0.18	0.494428	0.002967	0.491716	0.00297	−0.55162	0.109035	0.01994	−2.4E+18	5.84E+20
0.19	0.475099	0.002851	0.472493	0.002854	−0.5516	0.109053	0.01994	−3.2E+19	8E+21
0.2	0.456525	0.002739	0.454021	0.002742	−0.55158	0.109072	0.01994	−4.4E+20	1.1E+23
0.21	0.438678	0.002632	0.436271	0.002635	−0.55156	0.10909	0.01994	−6.1E+21	1.5E+24
0.22	0.421528	0.002529	0.419216	0.002532	−0.55155	0.109109	0.01994	−8.3E+22	2.06E+25
0.23	0.405048	0.00243	0.402827	0.002433	−0.55153	0.109127	0.01994	−1.1E+24	2.83E+26
0.24	0.389213	0.002335	0.387078	0.002338	−0.55151	0.109146	0.01994	−1.6E+25	3.88E+27
0.25	0.373997	0.002244	0.371946	0.002246	−0.55149	0.109164	0.01994	−2.1E+26	5.31E+28

Rewriting this and setting up an integral results in the following:

$$dV = f(V, t)dt,$$

$$\int_1^2 dV = \int_1^2 f(V, t)dt.$$

We can perform the integration on the right side *only* if we know the *t* dependence of *V*. Otherwise, we cannot do the integration, which is the usual case. In our standard example, we had $f(V,t) = 8.80 - 1.37 \times 10^{-3} * V^2$ where $V(t)$ was unknown. Hence we cannot simply integrate most of the differential equations we encounter. We must use one of the ordinary differential equation solvers.

■ 8.14 Program Step Control—Trapping

In using a VBA program to solve a differential equation, input would typically include the initial time t_0, the final time t_f, and the time step dt. Because of round-off, advancing from t_0 to t_f in increments of dt probably would not result in a value of t exactly equal to t_f. Instead we would be within a small fraction of a step size of t_f. We could run our program by putting the test inside a `Do While` loop as:

```
Do While abs(t - tf) > dt/100
      . . . .
      . . . .
      . . . .
      . . . .
Loop
```

Within the loop, `t` would be advanced according to the method used. At the beginning of each step, if `t` and `tf` were arbitrarily close to one another the solution steps would be ended. (The value of 100 is arbitrary.) This process of requiring two values to be arbitrarily close to one another without being identical is called *trapping*.

Similarly, the decision to print results at specified intervals could be controlled with a trap. Suppose we wish to print at a time interval of `pf`. After echoing the initial values, our next print, `np = t + pf`. At the end of each step we can test to see if we are arbitrarily close to `np` as:

```
If abs(t - np) <= dt/100 then
      . . . .
      . . . .
      . . . .
      np = np + pf
Endif
```

The statement `np = np + pf` advances the location of the next print. We cannot simply change `pf` if we wish to maintain a constant print frequency.

■ 8.15 Chapter 8 Exercises

8.1 Consider the equation $\dfrac{dz}{dx} = -z^2$. Given $z(0) = 1$ and $\Delta x = 0.25$, compute $z(0.25)$ and $z(0.5)$ using the Euler method. Compare to the exact solution,

$$z = \frac{1}{1 - x}.$$

8.2 Solve Exercise 8.1 using the improved Euler method. Compare to the exact solution given in Problem 8.1.

8.3 Solve Exercise 8.1 using the fourth-order Runge–Kutta method. Compare to the exact solution given in Problem 8.1.

8.4 Consider the following second-order differential equation:

$$\frac{d^2y}{dt^2} = -(y * \frac{dy}{dt})^2.$$

  a. Convert this to a pair of first-order equations.

  b. Given $y(0) = 1$, $y'(0) = 0.5$, and $\Delta t = 0.25$, calculate y and y' at $t = 0.25$ and 0.5 using Euler's method.

  c. Repeat **b.** using the improved Euler Method.

8.5 **a.** Convert the second-order differential equation given in Exercise 8.4 to a pair of first-order equations.

  b. Given $y(0) = 1$, $y'(0) = 0.5$, and $\Delta t = 0.25$, calculate y and y' at $t = 0.25$ and 0.5 using the fourth-order Runge–Kutta method.

  c. Use the Collatz criterion to comment on the step size used.

8.6 **a.** Write a generic VBA program to apply the fourth-order Runge–Kutta method to a pair of first-order equations. Use functions for the derivatives. To provide generality, these should be in the form $f(x,y,t)$ and $g(x,y,t)$ even if three arguments are not needed in a particular application (x is general; it may be a first derivative). Input should include the initial values of x, y, and t, the final value of t, the step size, and the print frequency. Provide labeled output of your input and results. Control step progress within a `Do While` loop that includes printing using "traps."

  b. Convert the second-order differential equation in Exercise 8.4 to a pair of first-order equations.

  c. Verify that your program is correct by applying it to the conditions and step size of Exercise 8.5b and comparing results.

d. Now apply your program with $\Delta t = 0.05$ to calculate y and y' out to $t = 5.0$. Output results at intervals of 0.1 secs. Also, compute and output the Collatz criterion at these intervals. What do you conclude about the step size?

8.7 The motion of a simple pendulum is described by the following second-order differential equation:

$$\frac{d^2\theta}{dt^2} = -9.81 \sin(\theta)$$

where θ is in radians.

a. Convert this to a pair of first-order equations.

b. Given $\theta(0) = 1.0$ and $\dfrac{d\theta}{dt}(0) = 0$, calculate θ and $\dfrac{d\theta}{dt}$ at $t = 0.5$ and 1.0 using the Euler method with a step size of 0.5.

8.8 a. Convert the second-order differential equation in Exercise 8.7 to a pair of first-order equations.

b. Given $\theta(0) = 1.0$ and $\dfrac{d\theta}{dt}(0) = 0$, calculate θ and $\dfrac{d\theta}{dt}$ at $t = 0.5$ and 1.0 using the fourth-order Runge–Kutta method with a full step size of 0.5.

c. Comment on the step size by computing the Collatz criterion.

8.9 a. Convert the second-order differential equation given in Exercise 8.7 to a pair of first-order equations.

b. Apply a generic VBA program to solve this equation pair using the fourth-order Runge–Kutta method following the instructions in Exercise 8.6.

c. Verify that your program is correct by applying it to the conditions and step size of Exercise 8.8b and comparing results.

d. Then apply your program to the equations of Exercise 8.7 to calculate θ and $\dfrac{d\theta}{dt}$ with a step size of 0.1 out to $t = 5.0$. Output results at t intervals of 0.5 secs. Also compute and output the Collatz criterion at these intervals. What do you conclude about the step size?

8.10 Consider the second-order differential equation:

$$y'' + 2(y')^2 - 3 * y * t = 0,$$

where y' is the first derivative of y with respect to time and y'' is the second derivative.

a. Convert this second-order equation into a pair of first-order equations.

b. It is given that $t = 0$, $y = 0.0$, and $y' = 1.0$. Using a full step size of $\Delta t = 0.1$, determine y and y' at $t = 0.1$ and $t = 0.2$ secs with the fourth-order Runge–Kutta method. Also, compute the Collatz criterion for y and y^1 at $t = 0.1$ and 0.2. What do you conclude about the step size?

8.11 a. Convert the second-order differential equation given in Exercise 8.10.

b. Apply a generic VBA program to solve this equation pair using the fourth-order Runge–Kutta method following the instructions in Exercise 8.6.

c. Verify that your program is correct by applying it to the conditions and step size of Exercise 8.10b and comparing results.

d. Apply your program to this pair using a full step size of $\Delta t = 0.01$ to determine y and y' out to $t = 5.0$ secs. Output results at t intervals of 0.05 secs. Also, compute and output the Collatz criterion at these intervals. What do you conclude about the step size?

8.12 The motion of a damped spring-mass system is described by the equation

$$m\frac{d^2x}{dt} = c\frac{dx}{dt} + kx.$$

a. Convert this second-order differential equation into a pair of first-order equations.

b. Apply a generic VBA program to solve this equation pair using the fourth-order Runge–Kutta method following the instructions in Exercise 8.6.

c. If $m = 0.0500$ kg, $c = 0.10$ kg/s, and $k = 1.0$ N/m, $x(0) = 2$, $\dot{x}(0) = 0$, apply your programto this pair using a full step size of $\Delta t = 0.005$ to determine x and $\frac{dx}{dt}$ outto $t = 0.50$ secs. Output results at t intervals of 0.10 secs. Also, compute andoutput the Collatz criterion for x and $\frac{dx}{dt}$ at these intervals. What do youconclude about the step size?

8.13 The motion of a pair of masses connected by spring 1 to a wall and to each other by spring 2, with damping on each mass, is described by the pair of second-order equations as

$$m_1\ddot{x}_1 + (c_1 + c_2)\dot{x}_1 - c_2\dot{x}_2 + (k_1 + k_2)x_1 - k_2x_2 = 0,$$
$$m_2\ddot{x}_2 - c_2\dot{x}_1 + c_2\dot{x}_2 - k_2x_1 + k_2x_2 = 0.$$

a. Convert this pair of second-order differential equations into four first-order equations.

b. Write a program to solve this equation set using the fourth-order Runge–Kutta method following the instructions in Exercise 8.6. Use four functions for the four derivatives.

c. If $m_1 = 0.0500$ kg, $m_2 = 0.025$ kg, $c_1 = 0.10$ kg/s, $c_2 = 0.075$ kg/s, $k_1 = 1000$ N/m, and $k_2 = 1500$ N/m, apply your program to these equations using a full step size of $\Delta t = 0.0025$ to determine x_1, $\dot{x}_1$, x_2, and $\dot{x}_2$ out to $t = 0.50$ secs. Let $x_1(0) = 1$, $\dot{x}_1(0) = 0$, $x_2(0) = 1$, $x_2(0) = 0$, Output results at t intervals of 0.10. Also, compute and output the Collatz criterion for the variables at these intervals. What do you conclude about the step size?

8.14 Repeat Exercise 8.1 using the predictor–corrector method with two corrections. Estimate the error.

8.15 Repeat Exercise 8.4 using the predictor–corrector method with two corrections. Estimate the error.

8.16 Repeat Exercise 8.7 using the predictor–corrector method with two corrections. Estimate the error.

9 Sets of Linear Equations

■ 9.1 Lots of Linear Equations

A few linear equations are not hard to solve by hand. Indeed, this topic is often considered before a student even begins the study of algebra. But in science and engineering it is often necessary to solve many linear equations, perhaps thousands or even millions, simultaneously. These equations may be the result of approximating partial differential equations as sets of linear equations. Large numbers of such equations frequently are encountered in computational fluid dynamic (CFD) modeling. In Chapter 11 we shall see how partial differential equations can be approximated as linear equations.

Because we are interested in being able to solve large numbers of such equations, we want methods that are basically independent of the number of equations and can be easily taught to a computer. We shall aim our solution methodology toward computer, rather than manual (by hand), solution. In particular, we shall be very systematic and not take advantage of situations that an observant human might exploit in a by-hand solution. As usual, results will be presented in decimal form, rather than as fractions.

We shall look at three methods for solving linear equations, each with its own advantages. One easily yields the determinant of the coefficients, and another the inverse of the matrix. Two of these methods are direct; one is iterative.

■ 9.2 Gaussian Elimination with Partial Pivoting

Gaussian elimination is a direct method of solving sets of linear equations. It is straightforward and easy to apply.

Consider a set of linear equations of the form

$$a_{1,1} * x_1 + a_{1,2} * x_2 + a_{1,3} * x_3 = c_1,$$
$$a_{2,1} * x_1 + a_{2,2} * x_2 + a_{2,3} * x_3 = c_2,$$
$$a_{3,1} * x_1 + a_{3,2} * x_2 + a_{3,3} * x_3 = c_3.$$

The first subscript on the a's indicates the row; the second subscript, the column. For more unknowns we would need more equations. The number of equations is usually equal to the number of unknowns. There cannot be fewer than this if a unique solution is to be obtained. If the number of equations exceeds the number of unknowns, which might happen if the coefficients represent multiple attempts to measure something, the system is overdetermined. We shall consider this case in Section 9.7.

An upper-diagonal form of a set of linear equations contains terms only along and above the diagonal. If the preceding equations could be manipulated in some way to put them in upper-diagonal form, we would have

$$\widetilde{a}_{1,1} * x_1 + \widetilde{a}_{1,2} * x_2 + \widetilde{a}_{1,3} * x_3 = \widetilde{c}_1,$$
$$\widetilde{a}_{2,2} * x_2 + \widetilde{a}_{2,3} * x_3 = \widetilde{c}_2,$$
$$\widetilde{a}_{3,3} * x_3 = \widetilde{c}_3,$$

where the ~'s represent coefficients whose values have been changed, in general, by the manipulation. Now it is a simple matter to begin with the last (nth) row and solve it for $x_n = \widetilde{c}_n / \widetilde{a}_{n,n}$. Then we can move up to the previous row and solve it for x_{n-1}, and so on. Working from the bottom up, all the x_i's can be obtained, one new unknown per equation.

A simple procedure exists to obtain a zero in the first column of all rows except the first row. First, we multiply row 1 by a factor $m = a_{2,1}/a_{1,1}$ and subtract it from row 2, coefficient by coefficient. For example, the new coefficient, $\widetilde{a}_{2,1}$, becomes $a_{2,1} - a_{1,1} * a_{2,1}/a_{1,1}$, which is zero. For the next row, we can multiply row 1 by the factor $a_{3,1}/a_{1,1}$ and then subtract this row from row 3. The general multiplier for row r, column c is $m = a_{r,c}/a_{r,r}$. This process can be continued down the first column until all rows except the first contain zeros in the first column.

When we have finished the first column, we move to the second column and perform a similar process to yield zeros in that column, starting with the first row below the main diagonal.

Applied to all columns except the last, this procedure will yield the desired upper-diagonal form. But what if a diagonal element, $\widetilde{a}_{r,r}$, is zero? Then the multiplier, m, would require dividing by zero and the process would fail. To avoid this, we could require that $a_{r,r}$ be the largest coefficient in the column (in magnitude) and switch two rows if it is not. Clearly, if zero is the biggest coefficient in the column in magnitude, we must have all zero's, and something is wrong with our equation set. The switching of rows like this is a technique called *partial pivoting*. Though not obvious, it also has the advantage that it reduces the round-off error that is introduced when we multiply the row containing the diagonal by a factor that is less than 1.

Pivoting does not simplify the Gaussian elimination solution. The pivoting just described is called partial because we are switching only rows to get the largest coefficient in the column under consideration on the diagonal. If we also considered columns in the switching process, we would be employing *full pivoting*. This is not often done in practice.

Let us apply Gaussian elimination to the following set of three equations and three unknowns:

$0x_1$	$-2x_2$	$+5x_3$	$= 12$
x_1	$-6x_2$	$+2x_3$	$= 9$
$4x_1$	$-1x_2$	$+1x_3$	$= 4$

Inspection of the first column shows that the 4 in row 3 is the largest element in magnitude in that column. Therefore, we shall switch (pivot) rows 1 and 3. Row 2 stays put. Pivoting only involves two rows at a time. After pivoting, we have

$4x_1$	$-1x_2$	$+1x_3$	$= 4$
x_1	$-6x_2$	$+2x_3$	$= 9$
$0x_1$	$-2x_2$	$+5x_3$	$= 12$

To eliminate the x_1 term in the second row, we can multiply row 1 by $m = \dfrac{a_{2,1}}{a_{1,1}} = \dfrac{1}{4} = 0.250$ and subtract it from row 2. Since computers generally perform decimal arithmetic, we shall, too. Even by hand, decimals are preferred, because manipulating odd fractions can easily lead to mistakes. In the steps that follow, we have written only the coefficients, relying on the columns to indicate the x location. Multiplying row 1 by 0.250 and subtracting it from row 2 yields

	1	-6	$+2$	9
$-(0.250$ R1	1	-0.25	0.25	1)
	0	-5.75	1.75	8

The multiplier relating rows 1 and 3 $= \dfrac{a_{3,1}}{a_{1,1}} = \dfrac{0}{4} = 0$, so nothing actually has to be done for this row. When we collect terms in each row we now have

4	-1	1	4
0	-5.75	1.75	8
0	-2	5	12

We have pivoted once. If we had not pivoted, the factor we would have applied to the first row before subtracting it from the second would have been $\dfrac{1}{0}$, an impermissible division by zero. Pivoting prevented this by moving the row containing the zero so that zero would not be on the diagonal. This allowed us to proceed in a systematic way that is easy to program.

When we now look at the second column from the diagonal down, we see the largest coefficient, in magnitude, is already in row 2 on the diagonal. Therefore, no pivoting is necessary for this column.

Our multiplier becomes $\dfrac{a_{3,2}}{a_{2,2}} = \dfrac{-2}{-5.75} = 0.348$. Then row $3 - m$ * row 2 yields

	0	−2	+5	12
−(0.348 R2	0	−2	0.609	2.78
	0	0	4.39	9.22

And again, collecting terms in each row, we now have

4	−1	1.0	4
0	−5.75	1.75	8
0	0	4.39	9.22

Forward elimination is now finished, so we can solve for the x's, beginning with the nth (third) row. Accordingly, we obtain

$$x_3 = \frac{9.22}{4.39} = 2.10.$$

Using this result in row 2, we obtain

$$x_2 = \frac{[8 - 1.75 * 2.10]}{-5.75} = -.752$$

and

$$x_1 = \frac{[4 - 0.752 - 2.10]}{4} = 0.287.$$

Substituting these results back into the original equation would indicate their correctness, accurate to the three significant figures we carried through these calculations.

A bonus of Gaussian elimination is that we can obtain easily the value of the determinant of the original equations, if desired. With p the number of pivots performed, the determinant is $(-1)^p$ times the product of the new diagonal coefficients. For our three-equation set, this is

$$\text{Det} = (-1)^p * \tilde{a}_{1,1} * \tilde{a}_{2,2} * \tilde{a}_{3,3}$$

$$= (-1)^1 * 4 * (-5.75) * 4.39 = 100.97 \approx 101.$$

In Section 9.5.1, we shall see that 101 is the correct value.

A word about the order of pivoting: it is properly done only as a column is being considered. One must not look at the coefficients in column 2 and pivot before column 1 is finished, column 3 before column 2 is finished, and so forth. This is because the relative magnitudes of the coefficients change as we perform the manipulations that produce zeros in the active column.

An example where premature pivoting leads to the wrong rearrangement occurs in the following set of coefficients:

−10	1	−5	5
9	2	7	15
5	−3	1	10

The first column has the largest coefficient in magnitude on the diagonal, so no pivoting is needed for that column. Since $|-3| > 2$, we might be tempted to switch rows 2 and 3 immediately. However, if we do not prematurely pivot, when we get to the second column, the coefficients are as follows:

−10	1	−5	5
0	2.9	2.5	19.5
0	−2.5	−1.5	10.0

We see that no pivoting actually is needed when we work with column 2 because $2.9 > |-2.5|$.

■ 9.3 Gaussian–Seidel Iteration

An alternate approach to solving linear equations is to use an iterative method called Gaussian–Seidel iteration, in which we systematically solve for the values of each unknown. Each calculation is easier than those performed in Gaussian elimination, but we may have to do a lot of them. This method may or may not converge and it is definitely slower for small numbers of equations. But for large equation sets it may be faster than Gaussian elimination.

Consider the same set of equations as before:

$0x_1$	$-2x_2$	$+5x_3$	$= 12$
x_1	$-6x_2$	$+2x_3$	$= 9$
$4x_1$	$-1x_2$	$+1x_3$	$= 4$

Instead of pivoting, we shall rearrange these equations to try to get the largest co-efficient (in magnitude) *in each row* on the diagonal. This is not always possible, but it is in this situation if row 3 becomes row 1, and row 1 becomes row 3. Row 2 does not have to be moved but unlike pivoting, it could be—that is, rearrangement may involve moving more than two equations. It also differs from pivoting in that we are looking for the largest element in each *row* rather than in each *column* to be on the diagonal. Finally, it is only done once at the beginning of the process. We do not have to wait for column manipulations.

Later, we shall introduce a sufficient condition for convergence related to rearrangement, but in the meantime let us rearrange this set as planned—that is, row 1 becomes row 3 and row 3 becomes row 1.

$$
\begin{aligned}
4x_1 & \quad -1x_2 & \quad +1x_3 & \quad = 4 \\
x_1 & \quad -6x_2 & \quad +2x_3 & \quad = 9 \\
0x_1 & \quad -2x_2 & \quad +5x_3 & \quad = 12
\end{aligned}
$$

We shall do something whose effectiveness is not obvious. We shall solve each equation for the variable on the diagonal. Thus,

$$
x_1 = \frac{[4 + x_2 - x_3]}{4},
$$

$$
x_2 = \frac{[9 - x_1 - 2 * x_3]}{-6},
$$

and

$$
x_3 = \frac{[12 + 2 * x_2]}{5}.
$$

We must assume initial conditions for the three x's. Though the rearrangement step we performed is critical, the choice of starting values is not. Shortly we shall present some data on the success of solving these equations with widely differing starting values. For simplicity, let $x_1{}^0 = x_2{}^0 = x_3{}^0 = 0$. The subscript denotes the variable; the superscript, the iteration number. Superscript 0 is the starting value. Then performing one full step yields

$$
x_1{}^1 = \frac{[4 + 0 - 0]}{4} = 1,
$$

$$
x_2{}^1 = \frac{[9 - 1 - 2 * 0]}{-6} = -1.33,
$$

and

$$x_3{}^1 = \frac{[12 + 2 * (-1.33)]}{5} = 1.87.$$

We have just completed one Gaussian–Seidel iteration. Notice that in each instance we used the latest values we had: that is, in calculating $x_2{}^1$ we used $x_1{}^1$, and in calculating $x_3{}^1$ we used $x_2{}^1$. If we consistently had used the values from the previous iteration, (the starting values in this iteration), the process would be known as *Jacobi iteration*. That is generally slower to converge since we are not using the most up-to-date information. Think of the advantage when planning a picnic of using the weather forecast from a few hours ago compared to that from yesterday.

Now let us do a second iteration. This iteration yields

$$x_1{}^2 = \frac{[4 - 1.33 - 1.87]}{4} = 0.200,$$

$$x_2{}^2 = \frac{[9 - 0.200 - 2 * 1.87]}{-6} = -.844,$$

and

$$x_3{}^2 = \frac{[12 + 2 * (-0.844)]}{5} = 2.06.$$

Have we gone far enough? How do we know when to stop? Answer: we could compute the approximate relative error after each iteration and continue until all in a given iteration are less than a prespecified value. After two iterations with this set we have

$$\varepsilon_{a,1} = \left| \frac{[0.2 - 1]}{0.2} \right| * 100 = 400\%,$$

$$\varepsilon_{a,2} = \left| \frac{[-0.844 - (-1.33)]}{-0.844} \right| * 100 = 57.6\%,$$

and

$$\varepsilon_{a,3} = \left| \frac{[2.06 - 1.87]}{2.06} \right| * 100 = 9.36\%.$$

These results are not very good because of the relative errors, but if the process is continued carrying three significant figures each time, we obtain the following results:

Iteration	3	4	5	6	7
x_1	0.274	0.285	0.287	0.287	0.287
x_2	−0.768	−0.755	−0.753	−0.753	−0.752
x_3	2.09	2.10	2.10	2.10	2.10

We see that by the fifth iteration we are getting virtually no change in the answers through the third significant figure. If we seek accuracy to seven significant figures— that is, a tolerance equal to $0.5 \times 10^{(2-7)}$ % applied to all x's—further calculations would show that 12 iterations are required.

We started with the x's all initially equal to zero. The iterations required with this set for seven-significant-figure accuracy with different starting values follow:

Starting Value	Iterations for All $\varepsilon_a <= 0.5 \times 10^{-5}$
0	12
1	11
10	13
100	14
10^9	23

Clearly, the number of needed iterations hardly depends on the starting value. So do not waste time trying to pick good ones; it is not critical, as it might be in solving non-linear equations. Can you use different starting values for the different variables? Sure, if you have a reason for choosing them to be different. But again, it is not critical.

Before moving on, let us look at what happens if we do not rearrange the equations to place the largest coefficients in each row on the diagonal. Our original equation set was

$$
\begin{aligned}
0x_1 &- 2x_2 &+ 5x_3 &= 12 \\
x_1 &- 6x_2 &+ 2x_3 &= 9 \\
4x_1 &- 1x_2 &+ 1x_3 &= 4
\end{aligned}
$$

We still have to do a bit of rearrangement since we cannot solve the first equation for x_1 that is not present. So we shall switch rows 1 and 2 to obtain

$$
\begin{aligned}
x_1 &- 6x_2 &+ 2x_3 &= 9 \\
0x_1 &- 2x_2 &+ 5x_3 &= 12 \\
4x_1 &- 1x_2 &+ 1x_3 &= 4
\end{aligned}
$$

Then we have

$$x_1 = \frac{[9 + 6 * x_2 - 2 * x_3]}{1},$$

$$x_2 = \frac{[12 - 5 * x_3]}{-2},$$

and

$$x_3 = \frac{[4 - 4 * x_1 + x_2]}{1}.$$

If we again let

$$x_1{}^0 = x_2{}^0 = x_3{}^0 = 0,$$

we achieve the following:

$$x_1{}^1 = \frac{[9 + 6 * 0 - 2 * 0]}{1} = 9,$$

$$x_2{}^1 = \frac{[12 - 5 * 0]}{-2} = -6,$$

and

$$x_3{}^1 = \frac{[4 - 4 * 9 - 6]}{1} = -38.$$

The second round of iteration yields

$$x_1{}^2 = -103, \qquad x_2{}^2 = 89, \qquad x_3{}^2 = -505.$$

Without continuing any further, this system seems to be diverging. We shall now introduce the test for diagonal dominance and show that this arrangement does not satisfy it.

9.3.1 Diagonal Dominance Test

We begin with a set of linear equations of the form

$$a_{1,1} * x_1 + a_{1,2} * x_2 + a_{1,3} * x_3 = c_1$$
$$a_{2,1} * x_1 + a_{2,2} * x_2 + a_{2,3} * x_3 = c_2$$
$$a_{3,1} * x_1 + a_{3,2} * x_2 + a_{3,3} * x_3 = c_3.$$

A sufficient condition for Gaussian–Seidel convergence of a set of n equations is that, for $n - 1$ equations, the magnitude of the coefficient of the diagonal term in the row is at least equal to the sum of the magnitudes of the other coefficients in that row to the left of the equal sign. (The constant on the right does not matter.) And, for at least one of the equations, the diagonal coefficient is greater in magnitude than the sum of the magnitudes of the other coefficients in that row to the left of the equal sign. This is a sufficient condition but not necessary. It is possible for the system to converge under Gaussian–Seidel iteration even if it is not satisfied. But convergence is guaranteed if it is satisfied. This condition is called diagonal dominance. Fortunately, it frequently occurs in problems in science and engineering, making the Gaussian–Seidel method of practical importance.

Mathematically, our requirement is:

$$\text{for } n - 1 \text{ equations, } |a_{ii}| >= \sum_{j=1, j \neq i}^{n} |a_{ij}|,$$

$$\text{and, for at least one equation, } |a_{ii}| > \sum_{j=1, j \neq i}^{n} |a_{ij}|.$$

Note that we compare the sum of the magnitudes and not the magnitude of the sum. For example, suppose that for row 1 we had $a_{1,1} = 6$, $a_{1,2} = 7$, and $a_{1,3} = -6$. By correctly using the sum of the magnitudes, we would obtain $6 < |7| + |-6| = 1,3$ and conclude that the equation does not satisfy the test. But if we incorrectly calculated the magnitudes of the sum, we would get $6 > |7 - 6| = 1$ and think the row satisfies the condition.

A common mistake in applying the convergence test is merely to note that each row has a coefficient larger than any other in that row without comparing it to the sum of the other coefficients. This dominant coefficient, we note, must be on the diagonal.

Let us apply the test to the formulations of the equation set, first to the rearranged form that converged.

Recall that the set is as follows:

$$
\begin{aligned}
4x_1 &- 1x_2 + 1x_3 &= 4 \\
x_1 &- 6x_2 + 2x_3 &= 9 \\
0x_1 &- 2x_2 + 5x_3 &= 12
\end{aligned}
$$

For row 1 we have $|4| = 4 > |-1| + |1| = 2$,

for row 2 we have $|-6| = 6 > |1| + |2| = 3$,

and, for row 3, we have $|5| = 5 > |0| + |-2| = 2$.

Since all three rows have $>$'s (only one would be required if we had $=$'s in the other two rows), this system clearly satisfies the test and will converge, as we have seen.

But for the formulation that diverged:

$$\begin{aligned}
x_1 &- 6x_2 + 2x_3 &= 9 \\
0x_1 &- 2x_2 + 5x_3 &= 12 \\
4x_1 &- 1x_2 + 1x_3 &= 4,
\end{aligned}$$

we would obtain

$$\begin{aligned}
|1| &= 1 < |-6| + |2| = 8, \\
|-2| &= 2 < |5| = 5, \text{ and} \\
|1| &= 1 < |4| + |-1| = 5.
\end{aligned}$$

In none of the rows did the diagonal coefficient equal or exceed the sum of the magnitudes of the other coefficients in that row, and the Gaussian–Seidel solution diverged.

9.3.2 Residuals

Besides computing the approximate relative errors and monitoring their progress, a conclusive way of evaluating whether an iterative scheme has converged is to put the current values into the equations to see how well the equations are satisfied. It is a bit like verifying that you are using the right key selected from among many by inserting it into the lock to see if it opens. The residual of a linear equation may be calculated by subtracting the right side (the constant) from the left side. Usually the absolute value of the difference is computed. If the values are exact, the residual would be zero. In actual cases, its size will depend on the extent of convergence, and the residuals will be different for each equation.

For the original equations of the previous set (rearrangement does not matter in calculating residuals), the residuals for the converging set after three iterations would be as follows:

$$R1: |a_{1,1}x_1 + a_{1,2}x_2 + a_{1,3}x_3 - c_1| = |0.274 - 6*(-0.768)$$
$$+ 2*(2.09) - 9| = 0.062,$$
$$R2: |a_{2,1}x_1 + a_{2,2}x_2 + a_{2,3}x_3 - c_2| = |0*0.274 - 2*(-0.768)$$
$$+ 5*(2.09) - 12| = 0.014, \text{ and}$$
$$R3: |a_{3,1}x_1 + a_{3,2}x_2 + a_{3,3}x_3 - c_3| = |4*0.274 - (-0.768)$$
$$+ 2.09 - 4| = 0.046.$$

These residuals are not bad but clearly some work needs to be done. Continuing the solution through seven iterations yields $x_1 = 0.287$, $x_2 = -0.752$, and $x_3 = 2.10$, which is the same three-significant-figure result we obtained from Gaussian elimination. The residuals become:

$$R1: |a_{1,1}x_1 + a_{1,2}x_2 + a_{1,3}x_3 - c_1| = |0.287 - 6 * (-0.752)$$
$$+ 2 * (2.10) - 9| = 0.001,$$
$$R2: |a_{2,1}x_1 + a_{2,2}x_2 + a_{2,3}x_3 - c_2| = |0 * 0.287 - 2 * (-0.752)$$
$$+ 5 * (2.10) - 12| = 0.004, \text{ and}$$
$$R3: |a_{3,1}x_1 + a_{3,2}x_2 + a_{3,3}x_3 - c_3| = |4 * 0.287 - (-0.752)$$
$$+ 2.10 - 4| = 0.000.$$

These results are consistent with the three significant figures carried throughout the calculations.

9.3.3 Relaxation

Sometimes we encounter a set of linear (or other) equations that appears to be converging under iteration, but the convergence is either very slow or oscillatory, as shown in Figure 9.1.

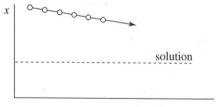

Slowly converging—
wish to speed up

Oscillating solution—
wish to add damping

I FIGURE 9.1 Slowly Converging and Oscillating Solutions.

In the top half of Figure 9.1, we would like to speed up its convergence. In the lower half, we would like to add damping to remove the overshoots in the convergence process. A method to do this is called relaxation. (We already mentioned a process called relaxation in connection with two-point boundary value problems in Chapter 8, but this is different.) The process will not make a diverging linear set converge, but it may make convergence faster. On the other hand, for nonlinear equations various relaxation schemes often are employed to coax along a solution that is not converging.

The basic idea is to take an iterated value and include some history in the solution by combining the tentative current value with its previous value by means of a relaxation factor. Selecting this factor is art, not science. Except in a few simple cases, the factor used is a matter of judgment, based on one's experience. Suppose you are planning to drive someplace where you have never driven before. You ask a friend how long it should take. Your friend says five hours. But then you remember that your friend drives a bit faster than you do, actually a lot faster, so you decide six hours is perhaps reasonable. You have just "relaxed" your friend's answer based on your judgment of your relative driving speeds. Of course it could go the other way, too. Your friend might have young children who can hardly pass a rest area or a fast-food place without stopping. So you might decide in those circumstances that four hours and fifteen minutes would be fine for you. You "relaxed" your estimate—that is, you applied a correction to it based on judgment.

Formally, we introduce a relaxation factor, λ, and apply it to the tentative and previous values to obtain the next value. The tentative value is what we would find with the conventional Gaussian–Seidel solution:

$$\tilde{x}_i^{\,n+1} = \frac{c_i - \sum\limits_{j=1, j \neq i}^{n} a_{i,j} * x_j}{a_{i,i}}.$$

Now let us apply the relaxation factor:

$$x_i^{\,n+1} = \lambda * \tilde{x}_i^{\,n+1} + (1 - \lambda) * x_i^{\,n}.$$

Note that if λ is greater than 1, $(1 - \lambda)$ will be negative. Let us apply relaxation to our rearranged set of equations, again starting with $x^0_1 = x^0_2 = x^0_3 = 0$ and with λ equal to 1.2. This is an arbitrary value. No claim is made that it will improve the rate of convergence in this case. Using the rearranged set of equations:

$$
\begin{aligned}
4x_1 & - 1x_2 & + 1x_3 & = 4 \\
x_1 & - 6x_2 & + 2x_3 & = 9 \\
0x_1 & - 2x_2 & + 5x_3 & = 12
\end{aligned}
$$

The tentative value is

$$\tilde{x}_1^{\,1} = \frac{[4 + x_2 - x_3]}{4} = \frac{[4 + 0 - 0]}{4} = 1.$$

Now we shall relax x_1 using $\lambda = 1.2$:

$$x_1^{\,1} = \lambda * \tilde{x}_1^{\,1} + (1 - \lambda) * x_1^{\,0} = 1.2 * 1 + (1 - 1.2) * 0 = 1.2.$$

The tentative value of x_2 is obtained and then relaxed as follows:

$$\tilde{x}_2 = \frac{[9 - x_1 - 2 * x_3]}{-6} = \frac{[9 - 1.2 - 0]}{-6} = -1.3, \text{ and}$$
$$x_2^{\,1} = \lambda * \tilde{x}_2^{\,1} + (1 - \lambda) * x_2^{\,0} = 1.2 * (-1.3) + (1 - 1.2) * 0 = -1.56.$$

Notice that we used the final value of x_1 and not its tentative value. The tentative value is discarded as soon as the final value of the variable for that iteration is determined. Our computation thus continues:

$$\tilde{x}_3 = \frac{[12 + 2 * x_2]}{5} = \frac{[12 + 2 * (-1.56)]}{5} = 1.78,$$

and

$$x_3^{\,1} = \lambda * \tilde{x}_3^{\,1} + (1 - \lambda) * x_3^{\,0} = 1.2 * (1.78) + (1 - 1.2) * 0 = 2.13.$$

It is important to note that we immediately relaxed each new tentative value after we computed it. The tentative value is used and quickly discarded. We do not wait until all the tentative values in an iteration have been computed before applying relaxation.

The second round of iteration would continue as follows:

$$\tilde{x}_1^{\,2} = \frac{[4 + x_2 - x_3]}{4} = \frac{[4 - 1.56 - 2.13]}{4} = 0.0775.$$

Now we shall relax it:

$$x_1^{\,2} = \lambda * \tilde{x}_1^{\,2} + (1 - \lambda) * x_1^{\,1} = 1.2 * 0.0775 + (1 - 1.2) * 1.2 = -0.147.$$

$x_2^{\,2}$ and $x_3^{\,2}$ would be determined in the same way.

Different values of λ may be used for each variable, and they can be changed from one iteration to the next if it seems appropriate. Changing values is based on experience, not the result of using a formula.

■ 9.4 Comparison of Gaussian Elimination and Gaussian–Seidel

As we have learned, Gaussian elimination is a direct method of solving sets of linear equations; Gaussian–Seidel is an iterative method. A single iteration is easy to perform, but we may have to perform many iterations in order to obtain a converged solution. How does the total effort compare?

Gaussian elimination without pivoting applied to n equations requires

$$\frac{4n^3 + 9n^2 - 7n}{6}$$

operations (see Dorn and McCracken, 1972). Gaussian–Seidel requires $2n^2 - n$ operations *per* operation or $(2n^2 - n) * $ It, where It is the number of iterations. If this is greater than the number of operations for Gaussian elimination, then we should use Gaussian elimination. For three equations, Gaussian elimination requires

$$\frac{4 * 3^3 + 9 * 3^2 - 7 * 3}{6} = 28$$

operations. Gaussian–Seidel would require $(2 * 3^2 - 3) * $ It $= 15 * $ It. Breakeven occurs when $28 = 15 * $ It, so convergence must be obtained in two iterations for Gaussian–Seidel to be competitive. Clearly this is unrealistic; Gaussian elimination is more efficient for small sets.

But what if $n = 100$? Gaussian elimination requires 681,550 operations while Gaussian–Seidel requires 19,900 * It operations. Breakeven occurs when $681,550 = 19,900 * $ It or when It $= 34$ iterations. This is possible, so Gaussian–Seidel might be competitive in this case. If we have a reasonable idea what the answers are, based on a physical understanding of the problem, our chances for a quick solution with Gaussian–Seidel improve.

■ 9.5 Matrix Inverses

It is trivial to solve a linear equation involving only one unknown—for example,

If $\qquad ax = c$

then $\qquad x = \dfrac{c}{a} = a^{-1}c.$

We cannot do this with sets of linear equations, but we can do something that follows this form, albeit in a more complicated way.

We may write a set of linear equations in matrix form as

$$[A][X] = [C], \tag{9.5.1}$$

where $[A]$ is an $n \times n$ matrix of the coefficients and $[X]$ and $[C]$ are column matrices with n elements each.

The matrix inverse $[A]^{-1}$ is defined as a matrix such that when it multiplies $[A]$ yields the identity matrix $[I]$ where $[I]$ is of the form

$$\begin{vmatrix} 1 & 0 & 0 \\ 0 & 1 & 0 \\ 0 & 0 & 1 \end{vmatrix}.$$

That is, $[I]$ has one's on the main diagonal and zero's elsewhere. Then

$$[A]^{-1}[A] = [I]. \tag{9.5.2}$$

If we were to premultiply both sides of Equation (9.5.1) by $[A]^{-1}$, we obtain

$$[A]^{-1}[A][X] = [A]^{-1}[C],$$

which, by applying (9.5.2) is

$$[I][X] = [A]^{-1}[C].$$

The identity matrix has the feature whereby multiplying a matrix by it does not change the matrix, just as multiplying a number by one does not change the number. Hence it is the matrix equivalent of one.

Thus,

$$[X] = [A]^{-1}[C]. \tag{9.5.3}$$

So all we have to do to find $[X]$ is premultiply $[C]$ by $[A]^{-1}$. But how do we find $[A]^{-1}$? It is not simply the reciprocal of $[A]$ in some fashion. First, we shall consider a method for finding a matrix inverse that is suitable for small matrices. Then we shall introduce another numerical technique that can be applied conveniently with a computer to find larger inverses.

9.5.1 Matrix Inverses by the Method of Cofactors

A *minor* of a matrix element is the determinant of the submatrix that results when the row and column of that element are excluded from the full matrix. A *cofactor* is a signed minor where the multiplying sign follows the pattern $(-1)^{(\text{row}+\text{column})}$. Thus, for row 1, column 1 we have $(-1)^{1+1} = (-1)^2 = 1$, but for row 1, column 2 we have $(1)^{(1+2)} = (-1)^3 = -1$. Alternately, we can view the premultiplying signs as alternating like the squares of a checkerboard:

+	−	+
−	+	−
+	−	+

It is to be emphasized that these signs multiply the minor and are *not* the ultimate sign of the result. That depends on both the premultiplier and the sign of the minor.

The value of a 2×2 determinant $\begin{vmatrix} a & b \\ c & d \end{vmatrix}$ is $a * d - c * b$.

Consider now the coefficients of our set of three equations:

0	−2	5
1	−6	2
4	−1	1

If we designate the cofactors of the $a_{i,j}$ as $A_{i,j}$, we obtain

$$A_{1,1} = + \begin{vmatrix} -6 & 2 \\ -1 & 1 \end{vmatrix} = +(-6 - (-2)) = -4, \text{ and}$$

$$A_{1,2} = (-) \begin{vmatrix} 1 & 2 \\ 4 & 1 \end{vmatrix} = (-)(1 - 8) = 7.$$

Continuing around the A matrix, we find that the remaining seven cofactors are

$$A_{1,3} = 23,$$
$$A_{2,1} = -3, \quad A_{2,2} = -20, \quad A_{2,3} = -8,$$
$$A_{3,1} = 26, \quad A_{3,2} = 5, \quad A_{3,3} = 2.$$

Now we need the determinant of the coefficient matrix. This we can get by multiplying any row or column times its cofactors. For example,

$$\text{Det} = a_{1,1} * A_{1,1} + a_{1,2} * A_{1,2} + a_{1,3} * A_{1,3}.$$

Since $a_{1,1} = 0$, this case is easier than usual. Thus we obtain

$$\text{Det} = 0 * (-4) - 2 * (7) + 5 * 23 = 101.$$

Now we have all the pieces we need to complete the calculation of the matrix inverse but we still must put them together correctly. The inverse is the *transpose* of the cofactor matrix with each element divided by the determinant. The transpose of a matrix is the original matrix with the rows and columns interchanged; that is, row 1 becomes column 1, row 2 becomes column 2, and so on. In equation form this is expressed as

$$[A]^{-1} = \frac{[cofactors]^T}{\text{Det}}.$$

In this case we obtain

$$[A]^{-1} = \begin{vmatrix} \dfrac{-4}{101} & \dfrac{-3}{101} & \dfrac{26}{101} \\[2mm] \dfrac{7}{101} & \dfrac{-20}{101} & \dfrac{5}{101} \\[2mm] \dfrac{23}{101} & \dfrac{-8}{101} & \dfrac{2}{101} \end{vmatrix},$$

and

$$[A]^{-1} = \begin{vmatrix} -0.0396 & -0.0297 & 0.257 \\ 0.0693 & -0.198 & 0.0495 \\ 0.228 & -0.0792 & 0.0198 \end{vmatrix}.$$

This was quite a bit of work. Let us pose and answer two questions. Is the result correct, and then, was it worth the effort?

We can prove our inverse is correct by multiplying it by the original matrix since $[A] * [A]^{-1} = [A]^{-1} * [A] = [I]$. To multiply two matrices, the number of rows of the first must equal the number of columns of the second. Since we have two 3 × 3 matrices, this condition obviously is satisfied. Let aa_{ij} be the elements of $[A]^{-1}$. A given element is in the answer,

$$d_{r,c} = R_r * C_c = a_{r,m} * aa_{m,c}.$$

That is, the row, r, of the first matrix is multiplied by column, c, of the second. The repeated double subscript in the coefficient product means it should be repeated over all possible values, in this case 1 to 3. Then

$$d_{r,c} = a_{r,1} * aa_{1,c} + a_{r,2} * aa_{2,c} + a_{r,3} * aa_{3,c}.$$

Thus, $d_{1,1} = a_{1,1} * aa_{1,1} + a_{1,2} * aa_{2,1} + a_{1,3} * aa_{3,1}$

$$= 0 * (-0.0396) + (-2) * (0.0693) + 5 * (0.228) = 1.001, \text{ and}$$

$$d_{2,1} = a_{2,1} * aa_{1,1} + a_{2,2} * aa_{2,1} + a_{2,3} * aa_{3,1}$$

$$= 1 * (-0.0396) + (-6) * (0.0693) + 2 * (0.228) = 0.0006.$$

Thus, to three-significant-figure accuracy, we have obtained the expected results for $d_{1,1}$ and $d_{2,1}$ of 1 and 0, respectively.

According to (9.5.3), we can solve the original equation set by multiplying the constant column by the inverse. In this case $[A]^{-1}$ is a 3×3 and $[C]$ is a 3×1. They are consistent. So $x_1 = $ row 1 of $[A]^{-1}$ times the column of $[C]$. We thus obtain

$$x_1 = aa_{1,1} * c_1 + aa_{1,2} * c_2 + aa_{1,3} * c_3 = -0.0396 * 12 - 0.0297 * 9$$

$$+ 0.257 * 4 = 0.286,$$

$$x_2 = aa_{2,1} * c_1 + aa_{2,2} * c_2 + aa_{2,3} * c_3 = 0.0693 * 12 - 0.198 * 9$$

$$+ 0.0495 * 4 = -0.752,$$

and

$$x_3 = aa_{3,1} * c_1 + aa_{3,2} * c_2 + aa_{3,3} * c_3 = 0.228 * 12 - 0.0792 * 9$$

$$+ 0.0198 * 4 = 2.10.$$

These are the same answers as before, so the process worked. But was it worth it? Not if all we wanted were the x's. But suppose we were doing a parametric study where the constants on the right-hand side represented applied voltages, wind loads, or heat loads, for example. If we now had to solve additional problems with the same coefficients but different loads, it would be easy to do this by repeating these multiplications using the new constant column vector. We would not have to go all the way back to the beginning, as we would if we were using Gaussian elimination or Gaussian–Seidel.

Suppose we now apply the inverse result to the same [A] but with [C] replaced by

1
2
3

Then $[X] = [A]^{-1}[C]$:

$$x_1 = aa_{1,1} * c_1 + aa_{1,2} * c_2 + aa_{1,3} * c_3 = -0.0396 * 1 - 0.0297 * 2$$
$$+ 0.257 * 3 = 0.672,$$
$$x_2 = aa_{2,1} * c_1 + aa_{2,2} * c_2 + aa_{2,3} * c_3 = 0.0693 * 1 - 0.198 * 2$$
$$+ 0.0495 * 3 = -0.178,$$

and

$$x_3 = aa_{3,1} * c_1 + aa_{3,2} * c_2 + aa_{3,3} * c_3 = 0.228 * 1 - 0.0792 * 2$$
$$+ 0.0198 * 3 = 0.129.$$

We have solved this problem with new constants simply by multiplying the inverse and the new column constants.

■ 9.6 Gaussian–Jordan Method

The Gaussian–Jordan method of solving linear equations is similar to Gaussian elimination except that rows are manipulated to obtain zeros for all off-diagonal elements, both above and below the main diagonal. This involves about twice as much work as Gaussian elimination. It is not competitive in effort unless we also need the inverse of the coefficient matrix. To obtain the inverse, we first append the identity matrix to the right side of the constant column and then perform Gaussian–Jordan with normalization of the diagonal elements. Normalization requires that we divide the coefficients in each row by the diagonal element in that row as we proceed. This produces ones on the main diagonal. The process will yield the solution vector where the constant column was, the inverse where the identity matrix was, and an identity matrix where the coefficient matrix was. Thus,

$$[A][C][I] + \text{Gaussian–Jordan} => [I][X][A]^{-1}.$$

We begin with the original coefficients, omitting the x labels.

0	-2	5	12
1	-6	2	9
4	-1	1	4

Now we append the identity matrix to the right side of the matrix, obtaining:

0	−2	5	12	1	0	0
1	−6	2	9	0	1	0
4	−1	1	4	0	0	1

Pivoting the expanded rows 1 and 3 gives us:

4	−1	1	4	0	0	1
1	−6	2	9	0	1	0
0	−2	5	12	1	0	0

We normalize row 1 by dividing it by 4:

1	−0.25	0.25	1	0	0	0.25
1	−6	2	9	0	1	0
0	−2	5	12	1	0	0

Subtracting row 1 from row 2 (row 3 already has a 0 at $a_{3,1}$) to obtain a 0 in position $a_{2,1}$ yields:

1	−0.25	0.25	1	0	0	0.25
0	−5.75	1.75	8	0	1	−0.25
0	−2	5	12	1	0	0

No new pivot is necessary since $|-5.75| > |-2|$. So we normalize row 2 by dividing it by -5.75, obtaining:

1	−0.25	0.25	1	0	0	0.25
0	1	−0.304	−1.39	0	−0.174	0.0435
0	−2	5	12	1	0	0

Subtracting (-0.25 * row 2) from row 1 and (-2 * row 2) from row 3 yields 0's in position $a_{1,2}$ and $a_{3,2}$:

1	0	0.174	0.652	0	−0.0435	0.261
0	1	−0.304	−1.39	0	−0.174	0.0435
0	0	4.39	9.22	1	−0.348	0.0870

Then we normalize row 3, dividing through each element by 4.39 yielding:

1	0	0.174	0.652	0	−0.0435	0.261
0	1	−0.304	−1.39	0	−0.174	0.0435
0	0	1	2.10	0.228	−0.0793	0.0198

Again, row 1 − 0.174 * row 3 and row 2 − (−0.304 * row3) yields:

1	0	0	0.287	−0.0397	−0.0297	0.258
0	1	0	−0.752	0.0693	−0.198	0.0495
0	0	1	2.10	0.228	−0.0793	0.0198

We have the identity matrix in the first three columns, the x's in column 4 (column $n + 1$) which agree with the values previously obtained with Gaussian elimination, and the inverse matrix in the last three columns (columns $n + 2$ to $2n + 1$), which likewise agrees with our previous result.

■ 9.7 Overdetermined Sets of Linear Equations

Occasionally, we may have more linear equations than we have unknowns. This could occur legitimately if our coefficients come from experimental measurement and we repeat the experiment. The data differ slightly, but we have no reason to reject some of the data in favor of the rest. How can we solve this overdetermined system?

Consider the following set of two equations:

$$2x_1 + 3x_2 = 13, \tag{9.7.1a}$$
$$4x_1 + 5x_2 = 23. \tag{9.7.1b}$$

It is easy to find that $x_1 = 2.0$ and $x_2 = 3.0$ satisfy this set. Suppose that an additional measurement yields the following coefficients, which are close to but not exactly the same as (9.7.1a):

$$1.99x_1 + 3.01x_2 = 12.95 \tag{9.7.1c}$$

Let us assume that the coefficients in (9.7.1c) are as likely to be valid as those in (9.7.1a). What should we do? One way to proceed would be to find the x's that minimize the residuals of the three equations where

$$R_i = a_{i,1}x_1 + a_{i,2}x_2 - c_i.$$

Since the residuals could be positive or negative and we want to avoid algebraic cancellation, we shall square each R_i and minimize the sum with respect to our variables, x_1 and x_2. Therefore,

$$R_1{}^2 + R_2{}^2 + R_3{}^2 = \text{sum}.$$

Differentiating this equation partially with respect to the two variables, x_1 and x_2, we find that

$$2R_1\frac{\delta R_1}{\delta x_1} + 2R_2\frac{\delta R_2}{\delta x_1} + 2R_3\frac{\delta R_3}{\delta x_1} = 0,$$

and

$$2R_1\frac{\delta R_1}{\delta x_2} + 2R_2\frac{\delta R_2}{\delta x_2} + 2R_3\frac{\delta R_3}{\delta x_2} = 0.$$

Now we have generated two new linear equations for the two unknowns x_1 and x_2. In our case, this yields

$$R_1 = 2x_1 + 3x_2 - 13,$$
$$R_2 = 4x_1 + 5x_2 - 23,$$

and

$$R_3 = 1.99x_1 + 3.01x_2 - 13.95.$$

Assuming four-significant-figure accuracy, and performing the differentiations and algebra, we obtain

$$23.96\,x_1 + 31.99\,x_2 = 143.8,$$

and

$$31.99\,x_1 + 43.06\,x_2 = 193.0.$$

Solving for x_1 and x_2, we have

$$x_1 = 2.148$$

and

$$x_2 = 2.886,$$

which is a little different from our original values of 2 and 3 but based on more, and equally valid, information.

We can generalize this process to any number of equations. Fortunately, the equations to be solved form an easily constructed pattern. If (c_i, c_j) represents the scalar product of the columns of each row of the original equations—for example, $(c_1, c_2) = a_{1,1} * a_{1,2} + a_{2,1} * a_{2,2} + a_{3,1} * a_{3,2}$, and so forth—the general form of the equations to be solved for the n x's is as follows:

$$(c_1,c_1)x_1 + (c_1,c_2)x_2 + \ldots (c_1,c_n)x_n = (a_1, b),$$
$$(c_2,c_1)x_1 + (c_2,c_2)x_2 + \ldots (c_2,c_n)x_n = (a_2, b),$$
$$\ldots$$
$$(c_n,c_1)x_1 + (c_n,c_2)x_2 + \ldots (c_n,c_n)x_n = (a_n, b).$$

More detail may be found in Scheid (1989).

■ 9.8 Excel Inverses and Linear Equation Solutions

Excel has a built-in capability to perform matrix calculations, including finding an inverse of a given matrix. The Excel function for doing that is MINVERSE(range) where *range* is the range of cells that contains the coefficients of the matrix whose inverse is sought—for example, A1:C3. This matrix must be square. Whenever a matrix function is executed in Excel, the "Control-Shift-Enter" keys must be pressed simultaneously. Pressing only "Enter" (which is tempting but wrong) will yield a number only in the upper left corner of the range where the result is to appear.

Excel does not have a function to solve a set of linear equations directly (which is surprising). Instead, we must solve the set by first obtaining the inverse and then multiplying the inverse times the constant column. We considered this process in Section 9.5. Symbolically, this is expressed as

$$[X] = [A]^{-1} * [C].$$

Suppose we wish to use Excel to solve the set considered earlier:

$0x_1$	$-2x_2$	$+5x_3$	$= 12$
x_1	$-6x_2$	$+2x_3$	$= 9$
$4x_1$	$-1x_2$	$+1x_3$	$= 4$

We could enter the coefficients, excluding the constant column, in, say, cells B3:D5 and the constants in F3:F5. Skipping a column is not necessary, but it improves the readability of the spreadsheet. To get the inverse of the coefficients on the left-hand side of the equations, we could sweep out a 3×3 range, say B8:D10, and in the left-hand corner cell of this range, B8, type =MINVERSE(B3:D5) and press "Control-

Shift-Enter" together. (Instead of typing in B3:D5, we could have swept over that range with the cursor after having opened the Matrix function, MINVERSE.) The inverse would then appear in B8:D10.

To complete the solution, we multiply the inverse times the constant column using MMULT(range1, range2) where MMULT is the Excel matrix multiplication routine; range1 is the range of the first matrix, the inverse in B8:D10; and range2 is the range of the second, the constant column in F3:F5. To do this we must highlight a new range of extent equal to the number of unknowns, say F8:F10, and in the first cell of this range, F8, type =MMULT(B8:D10,F3:F5) and press "Control-Shift-Enter." This places the X vector in cells F8:F10.

[A] (Coefficients)			[B] (Constants)
0	-2	5	12
1	-6	2	9
4	-1	1	4

[A]$^{-1}$ (Inverse of coefficients)			[X] X_1, X_2, X_3
-0.039604	-0.029703	0.257426	0.287129
0.069307	-0.019802	0.049505	-0.75248
0.227723	-0.079208	0.019802	2.09901

■ 9.9 Chapter 9 Exercises

9.1 Consider the equation set:

$$4x_1 + 8x_2 + 2x_3 = 2,$$
$$9x_1 + 5x_2 + x_3 = 6,$$
$$7x_1 + 3x_2 - 10x_3 = 7.$$

a. Use Gaussian elimination with partial pivoting to solve these equations.

b. Using your solution from **a.** calculate the determinant of the original matrix.

c. Solve Exercise 9.1 with Excel.

9.2 Consider the equation set in Exercise 9.1.

a. Show that this set satisfies the convergence criteria for the Gaussian–Seidel method of solving equations. Rearrange the equations as needed.

b. Assume starting values of $x_1^0 = x_2^0 = x_3^0 = 1$. Then perform three iterations of Gaussian–Seidel. Compute the approximate relative errors after each iteration.

    c. Using a relaxation factor of 1.25, perform one more iteration for the set.

    d. After the relaxed iteration, compute the residuals.

9.3 a. Find the inverse of the original matrix given in Exercise 9.1 using the method of cofactors.

    b. Prove that your inverse is correct by multiplying it times the original coefficient matrix.

    c. Solve the equation set using your inverse.

    d. Replace the right-hand side of the set in Exercise 9.1 by $c_1 = 12$, $c_2 = -16$, and $c_3 = 37$. Use your inverse to solve this set.

9.4 Consider the equation set:

$$0x_1 + 8x_2 + 2x_3 = 22,$$
$$x_1 + 5x_2 + 9x_3 = 26,$$
$$-10x_1 + 3x_2 + 7x_3 = 17.$$

    a. Use Gaussian elimination with partial pivoting to solve these equations.

    b. Using your solution from **a.**, calculate the determinant of the original matrix.

    c. Solve Exercise 9.4 with Excel.

9.5 Consider the equation set in Exercise 9.4.

    a. Show that this set satisfies the convergence criteria for Gaussian–Seidel. Rearrange the equations as needed.

    b. Assume starting values of $x_1{}^0 = x_2{}^0 = x_3{}^0 = 0$. Then perform three iterations of Gaussian–Seidel. Compute the approximate relative errors after each iteration.

    c. Using a relaxation factor of 0.70, perform one more iteration for the set.

    d. After the relaxed iteration, compute the residuals.

9.6 a. Find the inverse of the original matrix given in Exercise 9.4 using the method of cofactors.

    b. Prove that your inverse is correct by multiplying it times the original coefficient matrix.

    c. Solve the equation set using your inverse.

    d. Replace the right-hand side of the set in Exercise 9.4 by $c_1 = 10$, $c_2 = -20$, and $c_3 = 30$. Use your inverse to solve this set.

10 Arrays: Variables with a Family Name

■ 10.1 One-Dimensional Arrays

Frequently we work with sets of data whose members are all of the same kind. For example, we might be averaging test grades in a course. If there are 30 class members, it makes more sense to give a common name—for example, test3grade—to the grades and then subscript 30 numbers than it would to enter 30 different student names. In mathematics we might write these as $test3grade_1$, $test3grade_2$, $test3grade_3$, and so on. In VBA we would write them as `test3grade(1)`, `test3grade(2)`, `test3grade(3)`, and so on. Instead of a subscript, we have appended a number within parentheses to indicate which member of the test3grade set we are considering. In programming, the collection of the members of the set is called an *array*. If only one number appears in the parentheses, we have a one-dimensional array. If we are identifying the squares of a checkerboard or the coefficients of a set of linear equations, we would need two-dimensional arrays. A velocity field with x, y, and z components would require a three-dimensional array. In VBA, arrays of dimension up to 60 can be declared, a number not likely to be needed.

Before we can use arrays in a program we must tell the computer two things: (1) that the variable is an array and (2) how much storage to set aside for the members of that array. In VBA we do this with a `DIM` (short for "dimension") statement as:

```
DIM test3grade(30) as integer
```

This statement defines `test3grade` as an array with 31 (surprise!) locations for the variables with this common name (like a family last name). It is like making reservations where you inform the restaurant how many guests to expect in your party. Why 31? Because VBA will create the array and permit members numbered from 0 through 30, or 31 in all. In this respect it is unlike other higher-level languages such as C/C++ and FORTRAN. C/C++ would number the array members from 0 to 29; FORTRAN from 1 to 30. In both cases, each array would have 30 members. In our discussion we will refer to the nominal size of the array, excluding the first index value of zero (0).

Alternatively, we could specify only 30 members with `test3grade(1 to 30)`. This would not create the 0th member of the `test3grade` family, although the argument text is longer. But this format could be used to create negative index values—for example, `test3grade(-10 to 20)`—if this were desired.

We can also exclude a 0 index from all arrays used if we specify `Option Base 1` ahead of the Sub procedure in which `Dim` is included. Thus, we could have

```
Option Base 1
Sub Prog1()
Dim test3grade(30) as integer
```

It is important that we dimension an array. If we forgot to do this and tried to use an expression like `test3grade(22)`, VBA would think this is a subprogram named `test3grade` with one argument. Since this subprogram would not be defined, we would receive an error message.

It is also not acceptable to use the same name for both an array and a single variable. If we wanted to work with a single value of the `test3grade` array, we might use something like `test3 = test3grade(i)` where i had been defined. If we have more than one array in our program, they can be dimensioned on the same line as:

```
Dim A(10) as double, B(20) as double, C(40) as integer
```

The dimensions of each can be different, but the type of each must be declared.

Setting the size (or dimension) of `test3grade` at 31 does not mean we must actually use that many members in this array. We could have fewer—say, 6—but we must not have more, say, 35. If we were to exceed the size of the array, the extra variables might spill into space intended for other variables, causing some unpleasant surprises. So the dimension must represent an upper limit on the expected size of the array. On the other hand, it does not make sense to set the array size at 10,000 if it is never going to be larger than 100. You do not rent a banquet hall for a dinner for four people. In Section 10.3, we will discuss dynamic dimensioning, where the size of the array can be redefined before using, based on an input variable.

Individual array members are identified by the appropriate index inside parentheses, such as `test3grade(3)` or `test3grade(11)`. The index should be an integer.

Because array variables are indexed with integers, using `For` loops in conjunction with arrays is a powerful combination when working with large data sets. We shall see this in the example that follows.

■ 10.2 Example: Grade Averaging

Suppose the grades from a test are stored in the first column of a spreadsheet. The number of grades is in cell(1,1) with the grades listed below it. We want to write a VBA program to (1) input these grades, (2) compute the average grade, and (3) identify the highest and lowest grades. We shall do this with an array for the grades.

```
Sub grademanip ( )
Dim grade(50) as integer, n as integer, sum as integer, high as
    integer
Dim low as integer, i as integer, average as double
n = cells(1,1).value
sum = 0
high = 0
low = 100
For i = 1 to n
    grade(i) = cells(i+1,1).value
    sum = sum + grade(i)
    if grade(i) > high then
        high = grade(i)
    endif
    if grade(i) < low then
        low = grade(i)
    endif
next i
average = sum/n
Msgbox "average" & average & " high " & high &
    " low " & low
End Sub
```

In the second line of this program we have arbitrarily set the nominal size of the array at 50. If we encounter a class larger than this, we would have to increase the size of the grade array.

After obtaining the number of grades from the input spreadsheet, we initialized sum, high, and low. We initialized high to 0 and low to 100 because these represent extreme values, which will immediately be changed when the first actual grade is encountered.

Our For loop will then consider all n grades. We read them in from the appropriate cells. Note that the cell row changes and is one more than i because cell(1,1) contained the value of n instead of a grade. We then compare each new grade entered with the current high and low grades, and we modify those values as needed.

Following the For loop, we easily compute the average by dividing the sum of the input grades by their number. We could have done the summing and the comparison in separate For loops, but this would be less efficient than doing it all in one loop. Again, this required the initialization of sum, high, and low.

10.2.1 Indefinite Input from Excel

If an unknown number of data are input from the spreadsheet, a Do While loop could be continued until no more data were found. The VBA command for this condition is Empty. Thus the loop would continue as long as the cell read is not empty, for example:

```
n = 0
Do While Cells(n + 1, 1) <> Empty
    n = n + 1
    grade(n) = Cells(n, 1).Value
Loop
```

Notice that n is being advanced and counted.

10.2.2 Indefinite Input from a File

If an unknown number of data are input from a file, reading the file is continued until the end of the file is reached. The command is EOF(*number*) where *number* is the file number. Before the end of the file is reached, EOF is False. Thus,` for grades stored in file #4:

```
n = 0
Do While EOF(4) = False
    n = n + 1
    input #4, grade(n)
Loop
Close #4
```

Again, n is being advanced and counted. When the end of the file is reached, VBA sets EOF equal to True.

This process of counting data is more complicated for multi-dimensional arrays. For example, there are $n*(n + 1) = n^2 + n$ coefficients for an augmented matrix of the coefficients of n linear equations. If the total coeffects read in and counted is nn, the value of n is obtained by solving the quadratic equation

$$n^2 + n - nn = 0$$

for a positive value of n. Then the input data must be properly assigned. The whole input process is doable but not obvious.

■ 10.3 Dynamic Dimensioning

In Section 10.2, the Dim statement set the grade array size to 50. The first input line established the actual number of grades. Alternatively, we might have declared grade as an array of type integer with Dim without declaring its size. After we determined the size, we could use ReDim to specify the actual size of the grade array as:

```
Sub grademanip
Dim grade() as int
n = cells(1,1).value
ReDim grade(n)
```

We could use dynamic dimensioning with an indefinite array size by inserting the line

```
ReDim grade(n)
```

after the line

```
n = n + 1
```

in the program segment in Section 10.2.1 for spreadsheet data input or in Section 10.2.2 for file data input.

■ 10.4 Example: Sorting an Array

Suppose we want to sort the n values of grade already in the computer, from high to low. We could accomplish this by finding the highest grade, then looking at the remaining set for the next highest, and so on. A program segment to do this could look like this:

```
For i = 1 to n-1
    For is = i + 1 to n
        If grade(is) > grade(i) then
            temp = grade(is)
            grade(is) = grade(i)
            grade(i) = temp
        End if
    Next is
Next i
```

We need only sort the largest n - 1 grades. The one left over will be the lowest.

■ 10.5 Multidimensional Arrays

Often it is necessary to consider data that are properly described by more than one subscript. For example, to describe the location of a city we can give its latitude and longitude—that is, its location on a grid or checkerboard. Or, if we were solving a set of two equations and two unknowns, we might write them as

$$a * x + b * y = c,$$
$$d * x + e * y = f,$$

or more generally as

$$a_{1,1} * x_1 + a_{1,2} * x_2 = c_1,$$
$$a_{2,1} * x_1 + a_{2,2} * x_2 = c_2.$$

This is more systematic, requires fewer distinct variable names, and is easily expanded to many more equations.

A double-subscripted variable like $a_{r,c}$ would be referred to in VBA as a(r,c). Dimensioning a would require a statement like

```
Dim a(25,26) as double
```

where a would be an array with nominally 25 rows and 26 columns (actually, 26 rows and 27 columns because of the 0th row and 0th column). These might be the coefficients of a set of linear equations, with the actual last column containing the constants to the right of the equals signs. Thus it would be an *augmented matrix*.

Suppose we want to enter these coefficients from a file named lineareqinp.txt. Within the subprogram we could have the following:

```
Dim a(25,26) as double
Open "f:lineareqinp.txt" for input as #4
Open "f:lineareqout.txt" for output as #7

Input #4, n
For r = 1 to n
    for c = 1 to n + 1
            Input #4, a(r,c)
    Next c
Next r
close #4
REM output coefficients to file
For r = 1 to n
    For c = 1 to n + 1
            Print #7, a(r,c);
    Next c
Print #7,
Next r
.....
```

Notice that we have closed the input file after the data have been read. We have then echoed these coefficients to an output file. The semicolon at the end of the first Print line tells VBA not to drop to a new line each time a value of a(r,c) is written. It will drop when the Print #7, line is encountered after all the columns in the row are printed. With nothing to print from this line, the printer merely drops to the next line. In this way we would have a line of output corresponding to the n equations. Eventually we would have to close file #7.

Higher dimensional arrays are possible. If we solved a flow problem in three-dimensions, we would use arrays with three indices. For example, the pressure array might be P(100,100,100), which means this array would have one million members. Three-dimensional arrays can occupy a lot of storage space and require a lot of time to evaluate. VBA allows up to 60 dimensions in an array, which would require 60 numbers in the argument list.

10.5.1 Arrays in Arguments

Arrays passed in arguments require no special treatment once they have been dimensioned in the defining `Sub`. For example, if we have

```
Dim a(25,26) as double
```

in a `Sub` that inputs or defines a, we may pass a in the arguments list as

```
Call Rearrange(n,a)
```

The corresponding defining statement for `Rearrange` would be

```
Sub Rearrange(n,a)
```

Within `Rearrange`, a would be recognized as an array and could be used accordingly. Again, it is vital that variables be listed in the same order in the calling and defining statement argument lists.

The same procedure applies to `Function` statements.

■ 10.6 Gaussian Elimination Flowchart

A flowchart for implementing Gaussian elimination is shown here in Figure 10.1.

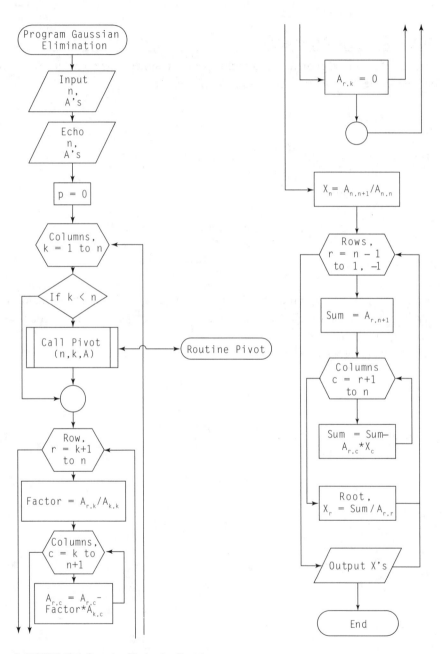

I FIGURE 10.1 Gaussian Elimination Flowchart.

10.6.1 Partial Pivoting Flowchart

A flowchart for partial pivoting is shown here in Figure 10.2.

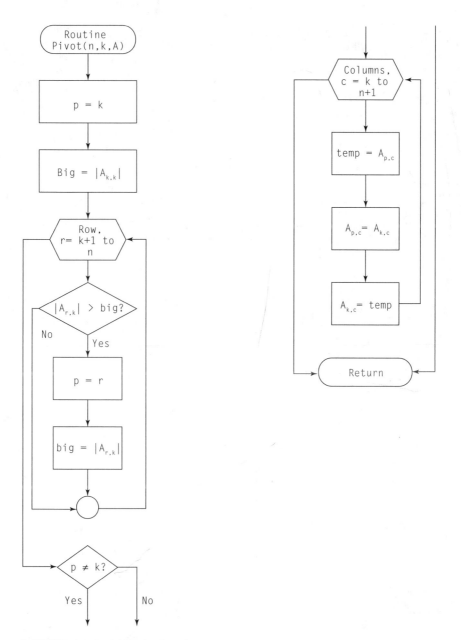

I FIGURE 10.2 Partial Pivoting Flowchart.

■ 10.7 Gaussian–Seidel Flowchart

A flowchart for Gaussian–Seidel is shown here in Figure 10.3.

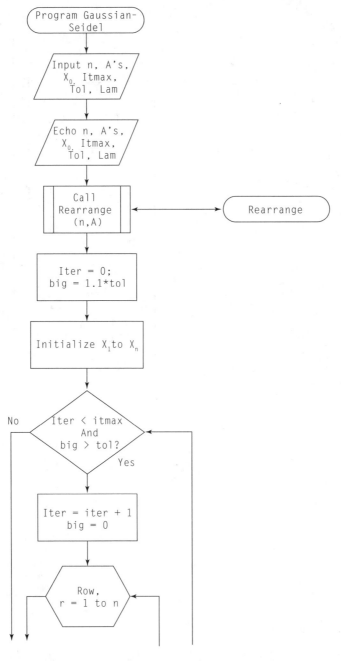

I FIGURE 10.3 Gaussian–Seidel Flowchart.

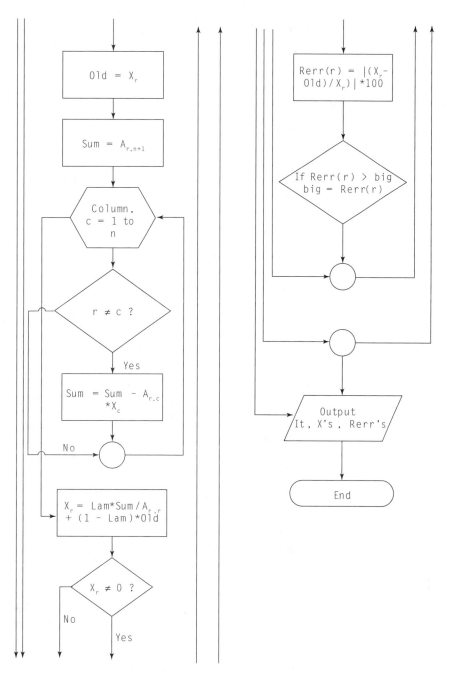

10.7.1 Rearrangement Flowchart

The flowchart in Figure 10.4 rearranges the equations for Gaussian–Seidel implementation.

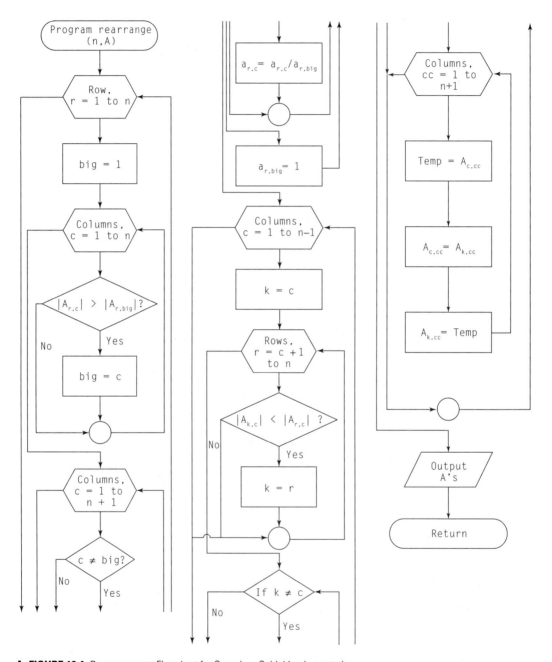

I FIGURE 10.4 Rearrangement Flowchart for Gaussian–Seidel Implementation.

■ 10.8 Gaussian–Jordan Flowchart

A flowchart for the Gaussian–Jordan method is shown here in Figure 10.5.

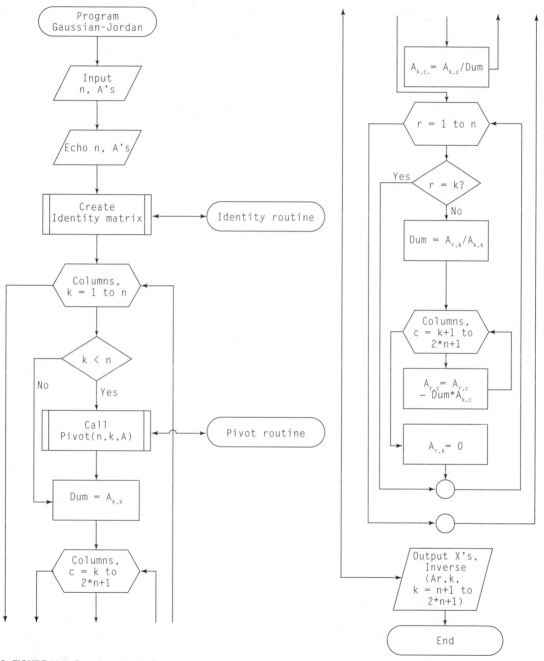

❙ FIGURE 10.5 Gaussian–Jordan Flowchart.

■ 10.9 Romberg Integration Flowchart

A flowchart for Romberg integration is shown here in Figure 10.6.

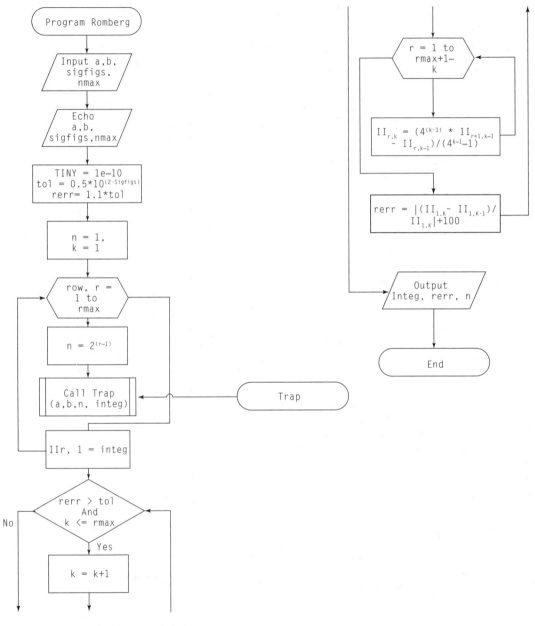

I FIGURE 10.6 Romberg Integration Flowchart.

■ 10.10 Thomas Algorithm Flowchart

A *tridiagonal matrix* is one that has nonzero terms only on the main diagonal and on the diagonals on either side of the main diagonal. Such matrices occur relatively

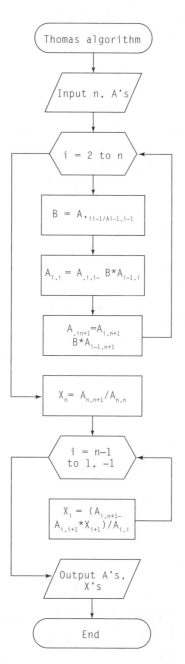

I FIGURE 10.7 Thomas Algorithm Implementation Flowchart.

frequently in engineering and science, often when one is numerically solving sets of partial differential equations.

The following matrix is tridiagonal:

$$
\begin{aligned}
1x_1 + 2x_2 \quad\quad\quad\quad &= 10 \\
3x_1 + 4x_2 + 5x \quad &= 20 \\
6x_2 + 7x_3 + 8x \quad &= 30 \\
9x_3 + 10x \quad &= 40
\end{aligned}
$$

Since at most three terms left of the equal sign will not be zero, Gaussian elimination can be modified so as to require fewer calculations. In fact, the number of operations can be reduced from being proportional to n^3 to being proportional to n (Jaluria and Torrance, 1986). The Thomas algorithm, or tridiagonal matrix algorithm, is one method of doing this. A flowchart for this method is shown in Figure 10.7.

■ 10.11 Chapter 10 Exercises (Use arrays when possible.)

10.1 Write a program to input the following grades. Use it to find the average, the high, and the low grades: 67, 80, 56, 75, 79, 62, 90, 76, 93, 68, 73, 91, 94.

10.2 The standard deviation of a population is given by the formula

$$
\sigma = \sqrt{\frac{\displaystyle\sum_{i=1}^{n}(x_i - \bar{x})^2}{n}}
$$

where $\bar{x}$ is the mean value. Write a program to compute the standard deviation of the grades in Exercise 10.1.

10.3 Write a program to solve a system of up to 10 linear equations using Gaussian elimination. Include a subroutine for partial pivoting. Follow the flowcharts provided in this chapter.

Apply your program to the following system:

$$
\begin{aligned}
4x_1 + 8x_2 + 2x_3 &= 2, \\
9x_1 + 5x_2 + x_3 &= 6, \text{ and} \\
7x_1 + 3x_2 - 10x_3 &= 7.
\end{aligned}
$$

Input the number of equations to be solved and the coefficients of the equations. Echo your input. Output the solution values and the determinant of the original coefficients calculated by your program.

10.4 Write a program to solve the system of equations in Exercise 10.3 using Gaussian–Seidel. Include a subprogram for rearrangement. Follow the flowcharts provided. Pick simple starting values. Apply (1) no relaxation, (2) overrelaxation, and (3) underrelaxation to your solution. Compare the number of iterations needed to obtain a solution with the largest approximate relative error less than 0.5×10^{-5} for the same starting values but different relaxation factors. Output this error and also the residuals of the original equations.

10.5 Write a program to solve the system of equations in Excerise 10.3 using Gaussian–Jordan with partial pivoting. Follow the flowcharts provided. Also calculate the inverse of the original coefficient matrix. Show that it is correct by multiplying the coefficient matrix times the inverse in your program. (This will require three nested For loops.)

10.6 Write a program to solve a system of up to 10 linear equations using Gaussian elimination. Include a subroutine for partial pivoting. Follow the flowcharts provided.

Apply your program to the following system:

$$x_1 + 2x_2 + 7x_3 + 2x_4 - x_5 = 23,$$
$$2x_1 + 11x_2 + x_3 - x_4 + 3x_5 = -5,$$
$$x_1 - x_2 - x_3 + 8x_4 - 2x_5 = -10$$
$$7x_1 + x_2 + x_3 + x_4 + 2x_5 = 19,$$
$$x_1 + 2x_2 + 3x_3 + 4x_4 - 15x_5 = 30.$$

Input the number of equations to be solved and the coefficients. Echo your input. Output the solution values and the determinant of the original coefficients calculated by your program.

10.7 Write a program to solve the system of equations in Exercise 10.6 using Gaussian–Seidel. Include a subprogram for rearrangement. Follow the flowcharts provided. Pick simple starting values. Apply (1) no relaxation, (2) overrelaxation, and (3) underrelaxation to your solution. Compare the number of iterations needed to obtain a solution with the largest approximate relative error less than 0.5×10^{-5} for the same starting values but different relaxation factors. Output this error and also the residuals of the original equations.

10.8 Solve the system of equations in Exercise 10.6 using Gaussian–Jordan with partial pivoting. Follow the flowcharts provided. Also calculate the inverse of the original coefficient matrix. Show that it is correct by multiplying the coefficient matrix times the inverse in your program. (This will require three nested For loops.)

10.9 Write a program to solve a system of up to 10 linear equations using Gaussian elimination. Include a subroutine for partial pivoting. Follow the flowcharts provided.

Apply your program to the following system:

$$x_1 + 2x_2 + 7x_3 + 2x_4 - x_5 = -5,$$

$$2x_1 + 11x_2 + x_3 - x_4 + 3x_5 = 23,$$

$$x_1 - x_2 - x_3 + 8x_4 - 2x_5 = -100,$$

$$7x_1 + x_2 + x_3 + x_4 + 2x_5 = 30,$$

$$x_1 + 2x_2 + 3x_3 + 4x_4 - 15x_5 = 19.$$

Input the number of equations to be solved and the coefficients. Echo your input. Output the solution values and the determinant of the original coefficients calculated by your program.

10.10 Write a program to solve the system of equations in Exercise 10.9 using Gaussian–Seidel. Include a subprogram for rearrangement. Follow the flowcharts provided. Pick simple starting values. Apply (1) no relaxation, (2) overrelaxation, and (3) underrelaxation to your solution. Compare the number of iterations needed to obtain a solution with the largest approximate relative error less than 0.5×10^{-5} for the same starting values but different relaxation factors. Output this error and also the residuals of the original equations.

10.11 Solve the system of equations in Exercise 10.9 using Gaussian–Jordan with partial pivoting. Follow the flowcharts provided. Also, calculate the inverse of the original coefficient matrix and show that it is correct by multiplying the coefficient matrix times the inverse. (This will require three nested For loops.)

10.12 Consider the following equations:

$$
\begin{aligned}
x_1 \quad -x_2 \qquad\qquad\qquad\qquad\qquad &= \quad 0 \\
0.4x_2 \qquad\qquad 0.5x_5 \qquad\qquad &= \quad 2.94 \\
-x_3 \qquad\qquad 0.98x_6 &= \quad 17.66 \\
x_4 \quad -x_5 \qquad\qquad &= \quad 29.43 \\
x_5 \quad 0.3x_6 &= \quad 58.86 \\
0.6x_2 \qquad\qquad\qquad 1x_6 &= \quad 0
\end{aligned}
$$

Use your Gaussian elimination program to solve these equations.

10.13 Use your Gaussian–Seidel program to solve the equations in Exercise 10.12.

10.14 Use your Gaussian–Jordan program to solve the equations in Exercise 10.12.

10.15 Write a program to implement the Romberg integration flowchart presented in Section 10.9. Apply your program to the example integral

$$\int_1^5 x^2 \ln(x)dx.$$

Assume a tolerance of 0.5×10^{-6} for ε_a.

10.16 Apply your Romberg integration program to

$$\int_0^{\pi/2} e^x * \cos(x)dx,$$

where x is in radians.

Assume a tolerance of 0.5×10^{-6} for ε_a.

10.17 Legendre polynomials are sometimes used in applied mathematics. The first three Legendre polynomials are as follows:

$$P_0(x) = 1, P_1(x) = x, \text{ and } P_2(x) = \frac{1}{2}(3x^2 - 1).$$

For $i >= 2$, subsequent Legendre polynomials may be generated by the following formula:

$$P_i = \frac{(2 * i - 1) * x * P_{i-1} - (i - 1) * P_{i-2}}{i}.$$

Write a program to generate Legendre polynomials. Let $x = 0.6$. Generate P_2 through P_{10}.

11 Curve Fitting

■ 11.1 Introduction

Frequently, measurements of physical quantities are plotted. Though the data points are discrete, we may try to connect them with a smooth curve and then determine the equation that defines that curve. Sometimes doing so may result in the establishment of a physical law. For example, when strain is plotted against stress, Hooke's law is apparent in the region where the elastic behavior is linear.

Good curve fits not only can provide insight into the underlying nature of the phenomenon being observed but also provide a compact representation of the measured data. Surely it would be more convenient if, say, we had a simple formula for determining a phone number rather than having to consult a book of numbers for the population in a local area.

■ 11.2 Linear Interpolation

Linear interpolation connects known data with a straight line. Points on this line can then be used to predict values of the dependent variable for a value of the independent variable that falls between the endpoints. (If the value of the independent variable falls outside the known values, the process is called *extrapolation*. For example, using known stock market data like the Dow-Jones industrial average for two dates could be used to predict its future value. But extrapolation is less accurate than interpolation because it extends data beyond their known range.)

Suppose we have values of the specific heat of liquid water at two temperatures:

$$T = 0°C \quad Cp = 1.008 \text{ J/(kg} \cdot °C), \text{ and}$$
$$T = 40°C \quad Cp = 0.998 \text{ J/(kg} \cdot °C).$$

But we need Cp at 10°C.

If we construct a straight line (see Figure 11.1) connecting the two known points, we can use this line to locate $T = 10°C$ and the corresponding Cp on the line. The actual Cp behavior might not be linear, but if the two known points are not far apart, the linear approximation probably is acceptable.

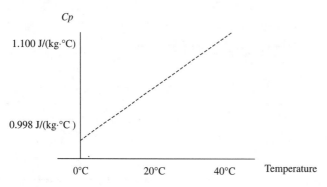

I FIGURE 11.1 Linear *Cp* Behavior.

Such calculations are easily visualized when tabulated, including the unknown value. Let us review the values:

Temperature	Cp
0	1.008
10	y
40	0.998

The unknown value of *Cp*, *y*, must bear the relation to the known *Cp*'s as $10°T$ does to the known temperatures. Thus,

$$\frac{10 - 0}{40 - 0} = \frac{y - 1.008}{1.008 - 0.998}.$$

That is,

$$y = 1.008 + \frac{(10 - 0) * (1.008 - 0.998)}{(40 - 0)}$$

$$y = 1.006.$$

If *Cp* had been desired at $T = 20°C$, we would have had

$$y = 1.008 + \frac{(20 - 0) * (1.008 - 0.998)}{(40 - 0)} = 1.003,$$

the average of 1.008 and 0.998, as expected.

■ 11.3 Lagrange Interpolating Polynomials

If more known points are to be utilized in interpolation, Lagrange polynomials might be used. The form for Lagrange interpolation involving three known data pairs is

$$f(x) = \frac{(x - x_1) * (x - x_2)}{(x_0 - x_1) * (x_0 - x_2)} * f(x_0) + \frac{(x - x_0) * (x - x_2)}{(x_1 - x_0) * (x_1 - x_2)} * f(x_1)$$
$$+ \frac{(x - x_0) * (x - x_1)}{(x_2 - x_0) * (x_2 - x_1)} * f(x_2).$$

For example, assume we have the following data:

$T = 0°C$ $Cp = 1.008$ J/(kg · °C)
$T = 20°C$ $Cp = 0.999$ J/(kg · °C)
$T = 40°C$ $Cp = 0.998$ J/(kg · °C)

Here, T assumes the role of x and Cp is $f(x)$. Then,

$$Cp(T) = \frac{(T - 20) * (T - 40)}{(0 - 20) * (0 - 40)} * 1.008 + \frac{(T - 0) * (T - 40)}{(20 - 0) * (20 - 40)} * 0.999$$
$$+ \frac{(T - 0) * (T - 20)}{(40 - 0) * (40 - 20)} * 0.998.$$

For $T = 10$, this yields $Cp(10) = 0.378 + 0.7493 - 0.1248 = 1.003$.

■ 11.4 Linear Regression

The simplest line to fit data is a straight line. Of course, many data sets do not correlate well with this fit. Nevertheless, it is the starting point for curve fitting, and constructing such a fit is called *linear regression*. With a little cleverness we can considerably extend its usefulness.

Such a curve is of the form

$$\bar{y} = a + b * x \qquad (11.4.1)$$

where we have added an overbar to indicate that y is the result of a curve fit rather than a measured value. Suppose we have a set of n measured stress, strain data pairs ($i = 1$ to n) as shown in Figure 11.2 where n is 5. We would like to pass the best straight line we can through these data. If we plotted the data on a graph by hand, this would involve moving a ruler around until an edge passed as close as possible to the majority of points.

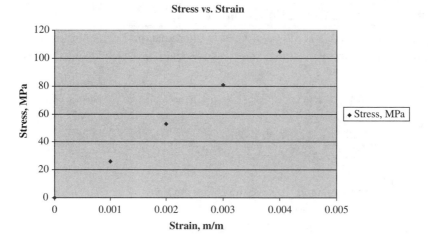

I FIGURE 11.2 Stress as a Function of Time.

How can we quantify "best straight line"? At each point we define the deviation, $\delta_i = y_i - \bar{y}_i$ where the unbarred y is the measured value we are approximating as $\bar{y}_i$ with our fit. Hence the deviation is defined analogously to the true error we have used often. The deviations represent the errors at each x_i, the amount by which our approximation misses the measured value at that point. Therefore, minimizing the sum of all the deviations seems a sensible goal. Some of the deviations, however, might be positive and some of them negative. These plus and minus values would tend to cancel each other and lead us to think our fit was better than it actually is. Hence, we shall instead (1) square each deviation producing only positive values, (2) sum these, and (3) seek to minimize this sum. It must be emphasized that we are computing the sum of the deviations squared:

$$\sum_{i=1}^{i=n} (\delta_i)^2,$$

not the square of the sum of the deviations,

$$\left[\sum_{i=1}^{i=n} \delta_i \right]^2.$$

This is an important difference.

Following Equation (11.4.1), at each i we have

$$\bar{y}_i = a + b * x_i. \tag{11.4.2}$$

Then the deviations, δ_i, will be $y_i - (a + b * x_i)$.

Squaring this and summing, we have

$$S_m = \sum_{i=1}^{i=n} [y_i - (a + b * x_i)]^2.$$

From calculus, we know that we can find a maximum or a minimum of a function by taking its derivative and setting it equal to zero. With what quantities should we take the derivative? Our unknowns, a and b. Thus we have two equations,

$$\frac{\delta S_m}{\delta a} = \frac{\delta \sum_{i=1}^{i=n} [y_i - (a + b * x_i)]^2}{\delta a} = -2 \sum_{i=1}^{i=n} [y_i - (a + b * x_i)] = 0, \quad (11.4.3)$$

and

$$\frac{\delta S_m}{\delta b} = \frac{\delta \sum_{i=1}^{i=n} [y_i - (a + b * x_i)]^2}{\delta b} = -2 \sum_{i=1}^{i=n} \{x_i * [y_i - (a + b * x_i)]\} = 0. \quad (11.4.4)$$

Since the right-hand sides of both equations are zero, the -2's can be divided out of each. We also can move the negative $(a + b * x_i)$ to the right-hand side of (11.4.3). Multiplying (11.4.4) through by the initial x_i and then moving the terms in parentheses to the right-hand side of the equation, we obtain for the two equations:

$$\sum_{i=1}^{i=n} y_i = \sum_{i=1}^{i=n} a + \sum_{i=1}^{i=n} b * x_i \qquad (11.4.5)$$

and

$$\sum_{i=1}^{i=n} x_i * y_i = \sum_{i=1}^{i=n} a * x_i + \sum_{i=1}^{i=n} b * x_i * x_i = \sum_{i=1}^{i=n} a * x_i + \sum_{i=1}^{i=n} b * x_i^2 \quad (11.4.6)$$

The first sum on the right-hand side of Equation (11.4.5) is simply a added n times or $n * a$. Making this simplification and reversing the left and right sides for convenience, our equations become:

$$a * n + b * \sum_{i=1}^{i=n} x_i = \sum_{i=1}^{i=n} y_i, \qquad (11.4.7a)$$

$$a * \sum_{i=1}^{i=n} x_i + b * \sum_{i=1}^{i=n} x_i^2 = \sum_{i=1}^{i=n} x_i * y_i. \qquad (11.4.7b)$$

Thus we have established two linear equations for the two unknowns, a and b. The equations are defined in terms of four sums of combinations of the measured data. Once these sums are calculated for a given set of data, the resulting equations can be solved in any convenient fashion.

11.4.1 Stress-Strain Curve

The following data were obtained for a certain metal:

x, Strain, m/m	y, Stress, MPa
0.0	0.0
0.001	26.0
0.002	53.0
0.003	81.0
0.004	105.0

When plotted, we obtain a nearly straight line, as shown in Figure 11.2.

Therefore, we shall attempt a linear fit to these data. We will need the sums of x and y and also x^2 and $x * y$. Placing these in tabular form we have:

	x	y	x2	x * y
	0.0	0.0	0.0	0.0
	0.001	26.0	0.000001	0.026
	0.002	53.0	0.000004	0.106
	0.003	81.0	0.000009	0.243
	0.004	105.0	0.000016	0.420
Sum	0.010	265.0	0.000030	0.795

Then Equations (11.4.7a) and (11.4.7b) become:

$$5a + 0.010b = 265.0$$
$$0.010a + 0.000030b = 0.795.$$

Solving these, we find $a = 0$ and $b = 26,500$.

Then our equation for stress, y, in terms of strain, x, is

$$y = 26,500x. \qquad (11.4.8)$$

11.4.2 Regression Coefficient

If we evaluate the stress at each of the given strains, we find the following values and deviations squared:

Strain	stress, meas.	Stress, pred (11.4.8)	deviation	dev.^{2}
0.0	0.0	0.0	0.0	0.0
0.001	26.0	26.5	0.5	0.25
0.002	53.0	53.0	0.0	0.0
0.003	81.0	79.5	−1.5	2.25
0.004	105.0	106.0	1.0	1.00
			Sum	3.50

The sum of the deviations squared is a good measure of the accuracy of the fit and can be used to compare the goodness of different assumed forms.

Another way of comparing the goodness of fit is to compute the regression coefficient, r. A general form for this expression is

$$r = \sqrt{\frac{S_t - S_r}{S_t}} \tag{11.4.9}$$

where

$$S_t = \sum_{i=1}^{i=n} (y_i - \bar{y})^2,$$

that is, the sum of the squares of the differences between the y's and the mean value of y and S_r is the sum of the deviations squared. For our data,

$$\bar{y} = \sum_{i=1}^{i=5} \frac{y_i}{5} = \frac{105}{5} = 21.0.$$

Then $S_t = (0 - 21)^2 + (26 - 21)^2 + (53 - 21)^2 + (81 - 21)^2 + (105 - 21)^2 = 12{,}146$ and $S_r = 3.5$, as we have seen.

Then $r = \sqrt{\dfrac{12{,}146 - 3.5}{12{,}146}} = 0.9999.$

Since this is extremely close to 1.0, the fit we have is very good, as we would expect in the linear region of a stress-strain curve.

This form of the regression coefficient is more general than the form often given and can be applied to nonlinear correlations involving one independent variable, as well.

Be careful in the interpretation of correlation coefficients. Some authors report r^2 instead of r. In general, r's in the physical sciences and technology are better than those in the social sciences where the lack of control results in greater variation in data.

11.4.3 The Independent Variable?

The choice of x or y as the independent variable is up to the analyst, based on the physics of the process being modeled. Mathematics will not dictate this choice.

Instead of $\bar{y} = a + b * x$, we might have chosen

$$\bar{x} = d + c * y.$$

But if we adopt this form, we must begin by summing y, y^2, x, and $x * y$ and perform a complete regression analysis. It is not enough to rearrange Equation (11.4.1) to solve for x. Regression is a statistical process, involving numbers raised to powers; it is not simply an exercise in algebra.

■ 11.5 Polynomial Regression

Although a straight line nicely fits the stress-strain data, frequently more sophisticated fits are needed. Consider the following data for the specific heat of water between 0 and 100°C and their plot:

T, °C	Cp, kJ/(kg · °C)
0	1.008
20	0.999
40	0.998
60	1.000
80	1.003
100	1.007

These data are plotted in Figure 11.3.

Clearly, the data here will not be well-represented by a straight line. If a polynomial fit of the form

$$\bar{y}_i = a + b * x_i + c * x_i^2 + d * x_i^3 + \ldots + z * x_i^n \tag{11.5.1}$$

is assumed, and we again minimize the sum of the deviations squared by taking derivatives of that sum with respect to each of the undetermined coefficients, the following set of equations results:

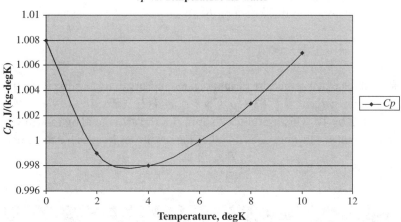

Cp vs. Temperature for water

I FIGURE 11.3 *Cp* of Water as a Function of Temperature.

$$a*n \quad + b*\sum_{i=1}^{i=n} x_i + c*\sum_{i=1}^{i=n} x_i^2 + d*\sum_{i=1}^{i=n} x_i^3 + \sum_{i=1}^{i=n} \ldots = \sum_{i=1}^{i=n} y_i,$$

$$a*\sum_{i=1}^{i=n} x_i + b*\sum_{i=1}^{i=n} x_i^2 + c*\sum_{i=1}^{i=n} x_i^3 + d*\sum_{i=1}^{i=n} x_i^4 + \sum_{i=1}^{i=n} \ldots = \sum_{i=1}^{i=n} x_i*y_i,$$

$$a*\sum_{i=1}^{i=n} x_i^2 + b*\sum_{i=1}^{i=n} x_i^3 + c*\sum_{i=1}^{i=n} x_i^4 + d*\sum_{i=1}^{i=n} x_i^5 + \sum_{i=1}^{i=n} \ldots$$
$$= \sum_{i=1}^{i=n} x_i^2*y_i,$$

$$a*\sum_{i=1}^{i=n} x_i^3 + b*\sum_{i=1}^{i=n} x_i^4 + c*\sum_{i=1}^{i=n} x_i^5 + d*\sum_{i=1}^{i=n} x_i^6 + \sum_{i=1}^{i=n} \ldots$$
$$= \sum_{i=1}^{i=n} x_i^3*y_i. \tag{11.5.2}$$

Higher-order fits require more equations of this form. A quadratic fit, e.g., $y = a + bx + cx^2$ has 3 undetermined coefficients.

We must have as many equations as we have unknown coefficients and as many columns to the left of the equal sign as we have coefficients. Notice that each succeeding row looks as if we had simply multiplied through the previous row by x_i inside the summation. As a consequence, the x to a given power slides to the left as we move to successive equations. This pattern makes it easy to write out the needed equations for whatever degree we want. In practice, a polynomial of 5 or 6 is about the maximum to be considered. Additional terms become larger and larger (or smaller and smaller for x's less than one) because of the higher powers involved. The additional round-off negates the improvement of adding higher powers to the fitting equation.

The curve in the plot of Cp vs. T resembles a parabola. Therefore, we shall try a quadratic equation fit to these data of the form

$$Cp = a + b * T + c * T^2$$

and solve three equations of the form indicated in Equation (11.5.2). To do this, we will need the sums of $T, T^2, T^3, T^4, Cp, Cp * T$, and $Cp * T^2$. Before we do this we shall divide each T by 10. This useful trick simplifies the arithmetic by reducing the size of the sums we must carry but does not alter the results, as long as we remember that our fit will be for Cp as a function of $\dfrac{T}{10}$.

$\dfrac{T}{10}$	Cp	$\left(\dfrac{T}{10}\right)^2$	$\left(\dfrac{T}{10}\right)^3$	$\left(\dfrac{T}{10}\right)^4$	$Cp * \dfrac{T}{10}$	$Cp * \left(\dfrac{T}{10}\right)^2$
0	1.008	0	0	0	0	0
2	0.999	4	8	16	1.998	3.996
4	0.998	16	64	256	3.992	15.968
6	1	36	216	1296	6	36
8	1.003	64	512	4096	8.024	64.192
10	1.007	100	1000	10,000	10.07	100.7
Sum 30	6.015	220	1800	15,664	30.08	220.9

Thus our equation set becomes:

$$
\begin{aligned}
6a \quad + \quad 30b \quad + \quad 220c \quad &= 6.015, \\
30a \quad + \quad 220b \quad + \quad 1800c \quad &= 30.08, \\
220a \quad + \quad 1800b \quad + \quad 15{,}664c \quad &= 220.9.
\end{aligned}
$$

Solving these for a, b, and c, we obtain $a = 1.009$, $b = -0.004951$, and $c = 0.0005022$. Thus our equation for $\overline{C}p$ is

$$\overline{C}p = 1.009 - 0.004951 * \frac{T}{10} + 0.0005022 * \left(\frac{T}{10}\right)^2.$$

When we apply this fit to the measured data we obtain the following values, deviations squared, and differences from Cp_{avg}:

$\dfrac{T}{10}$	$\bar{C}p$	dev²	$(\bar{C}p - Cp_{avg})^2$	
0	1.009	1.0000E-06	4.2250 × E-5	
2	1.0011	4.4386E-06	1.8225E-06	1:9
4	0.9972	5.9105E-07	1.58859E-05	2.77
6	0.9974	6.9001E-06	1.39022E-05	2.63
8	1.0015	2.1527E-06	3.34745E-07	9.35
10	1.0097	7.3441E-06	2.98584E-05	5.1984
Sum 30		2.2427E-05	1.5115E-04	

Thus $S_t = 1.5115\text{E-}04$, $S_r = 2.2427\text{E-}05$, and

$$r = \sqrt{\frac{S_t - S_r}{S_t}} = \sqrt{\frac{1.5115\text{E-}04 - 2.2427\text{E-}05}{1.5115\text{E-}04}} = 0.923.$$

This is a good fit. A cubic fit would represent the actual behavior even better. We will see later that such a fit would lead to an r^2 value of 0.99!

■ 11.6 The Power Law

Many physical quantities are not well represented by a polynomial equation. Logarithmic and exponential fits come to mind immediately as alternatives. Fortunately, we can make use of what we already know of linear regression to develop the relevant equations. Other fits are possible as the result of two independent variables, or multiple regression. We shall consider those in Section 11.8.

If a plot of y vs. x does not yield a reasonable straight-line or polynomial fit, we might want to plot the data on log-log axes. Such axes follow a logarithmic rather than a linear scale. If we then obtain a straight line, we are dealing with a *power law* relation. Let us see why this is true.

If we obtain a straight line in this situation, we are really plotting log y as a function of log x. (It does not matter whether we are using common or natural logs, as long as we remember which we are using.) Then we have

$$\log(y) = A + b\log(x).$$

If we are using common logs, we can apply this equation to the base 10; that is,

$$10^{\log(y)} = 10^{(A + b\log(x))} \text{ where log is } \log_{10}.$$

The sum of the exponents can be written as the product of the base to those powers. Hence,

$$10^{\log(y)} = 10^A * 10^{b \log(x)}.$$

But the log of a number for a given base raised as a power of that base is just the number, so for base Ba, as an example, we have

$$Ba^{\log_{Ba}(x)} = x.$$

A number times a log of a quantity may be written as the log of that quantity raised to the number: that is,

$$b \log_{Ba}(x) = \log_{Ba}(x^b).$$

Combining these results and letting $10^A = a$, our expression can be written as

$$y = ax^b. \tag{11.6.1}$$

This common form is known as the power law. We can obtain a and b by replacing x and y in Equations (11.4.7a) and (11.4.7b) by log x and log y, respectively. Then the equation set to be solved is as follows:

$$A * n + b * \sum_{i=1}^{i=n} \log(x_i) = \sum_{i=1}^{i=n} \log(y_i),$$

$$A * \sum_{i=1}^{i=n} \log(x_i) + b * \sum_{i=1}^{i=n} \log((x_i))^2 = \sum_{i=1}^{i=n} \log(x_i) * \log(y_i).$$

11.6.1 Power Law Example

Let us see how the power law applies to an example. The following data relating pressure drop, ΔP, and flow velocity, V, were obtained in a fluid mechanics lab:

V	ΔP
2.5	28.0
5.0	98.0
7.5	203
10	341
12.5	509
15	707
17.5	933
20	1186

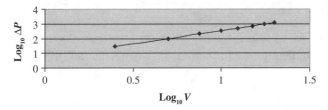

I FIGURE 11.4 Power Law Curve Fit.

A plot of the log, base 10 of ΔP as a function of the log, base 10 of V, yielded nearly a straight line as shown in Figure 11.4.

When log, base 10, was taken for all of the V and ΔP's, we obtained

	log(V)	log(ΔP)	(log(V))^2	log(V)*log(ΔP)
	0.3979	1.447	0.1584	0.5759
	0.6990	1.991	0.4886	1.392
	0.8751	2.306	0.7657	2.018
	1	2.533	1	2.533
	1.097	2.707	1.203	2.969
	1.176	2.849	1.383	3.351
	1.243	2.970	1.545	3.692
	1.301	3.074	1.693	3.999
Sum	7.789	19.877	8.237	20.530

The resulting equation set is

$$8A + 7.789b = 19.877,$$
$$7.789A + 8.237b = 20.53.$$

Solving for A and b, we obtain $A = 0.732$ and $b = 1.8$. Then

$$a = 10^A = 5.400,$$

so our curve fit is

$$\Delta P = 5.38 * V^{(1.80)}.$$

By calculating the differences between ΔP and ΔP_{avg} squared (S_t) and the sum of the deviations squared (S_r), we obtain $S_t = 111890$, and $S_r = 585.838$ for a regression coefficient value, r, of 0.997.

■ 11.7 The Exponential Fit

Suppose a plot of log y vs. x (note x is linear) gives the appearance of a straight line. If we (arbitrarily but conveniently) use the natural log, such data yield a line of the form

$$\ln(y) = A + bx.$$

If we apply this equation to the base, e, we have

$$e^{\ln(y)} = e^{(A+bx)} = e^A e^{bx}.$$

Again, the natural log of a number as the power of e is just the number, and if we let $e^A = a$, we obtain:

$$y = ae^{bx}. \tag{11.7.1}$$

We can obtain a and b by replacing y in Equations (11.4.7a) and (11.4.7b) by ln y. The equation set to be solved is:

$$A * n + b * \sum_{i=1}^{i=n} x_i = \sum_{i=1}^{i=n} \ln(y_i),$$

$$A * \sum_{i=1}^{i=n} x_i + b * \sum_{i=1}^{i=n} x_i^2 = \sum_{i=1}^{i=n} (x_i *) \ln(y_i).$$

Many natural phenomena follow such a law, including radioactive decay (b is then negative) and population behavior (positive or negative b).

11.7.1 Exponential Example

As an example, assume we have the following data:

t	y
1	1.92
2	2.97
3	4.26
4	4.71
6	7.03
8	8.00
10	11.0
12	12.3
14	15.6

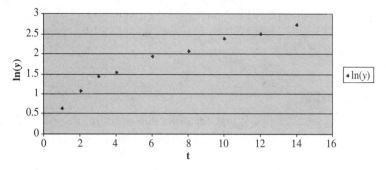

I FIGURE 11.5 Exponential Curve Fit.

When the ln of y is plotted against t as in Figure 11.5, the resulting shape is approximately linear, strongly suggesting an exponential fit to the data.

Computing $\ln(y)$ from y and performing regression analysis, we obtain the following values:

	t	y	$\ln(y)$	t^2	$\ln(y) * t$
	1	1.92	0.654	1	0.654
	2	2.97	1.09	4	2.18
	3	4.26	1.45	9	4.35
	4	4.71	1.55	16	6.2
	6	7.03	1.95	36	11.7
	8	8.00	2.08	64	16.64
	10	11.02	2.4	100	24
	12	12.30	2.51	144	30.12
	14	15.64	2.75	196	38.5
Sum	60		16.434	570	134.344

The resulting equation set is

$$9A + 60b = 16.34$$
$$60A + 570b = 134.3$$

When these equations are solved we obtain $A = 0.854$ and $b = 0.146$. Then $a = e^A = e^{(0.854)} = 2.35$ so our fit is

$$y = 2.35 * e^{(0.146t)}$$

Notice the exponential factor is negative, indicating decay.

By calculating the differences between y and y_{avg} squared (S_t) and the sum of the deviations squared (S_r) we obtain $S_t = 227.7$ and $S_r = 12.24$ for a regression coefficient value, r, of 0.973.

■ 11.8 Multiple Regression

Sometimes a dependent variable might be a function of more than one independent variable. For example, many thermodynamic properties are a function of both temperature and pressure.

Suppose we believe that variable y is a linear function of x and z in the form

$$y = a + bx + cz \tag{11.8.1}$$

The values of a, b, and c can be found by solving a set of three linear equations for the n data triplets (x, y, and z). Accordingly,

$$n * a + \sum_{i=1}^{i=n} b * x_i + \sum_{i=1}^{i=n} c * z_i = \sum_{i=1}^{i=n} y_i,$$

$$\sum_{i=1}^{i=n} a * x_i + \sum_{i=1}^{i=n} b * x_i^2 + \sum_{i=1}^{i=n} c * x_i * z_i = \sum_{i=1}^{i=n} x_i * y_i, \text{ and}$$

$$\sum_{i=1}^{i=n} a * z_i + \sum_{i=1}^{i=n} b * x_i * z_i + \sum_{i=1}^{i=n} c * z_i^2 = \sum_{i=1}^{i=n} z_i * y_i. \tag{11.8.2}$$

For a fit based on the product of two variables raised to a power, we could have

$$y = A * x^b * z^c$$

and employ $\log(x)$, $\log(y)$, and $\log(z)$ instead of x, y, and z in Equation (11.8.2).

11.8.1 Multiple Regression Example

In a certain heat transfer measurement, the Reynolds number, Re; a nondimensional length, L; and the Nusselt number, Nu, were measured. Examination of the results indicated that Nu appeared to be a function of both L and Re. This was a matter of judgment. It was proposed that a fit of the form $Nu = aL^b Re^c$ be attempted.

As shown in Table 11.1, this is also a function with two independent variables, but it is not linear as is Equation (11.8.1). However, as we saw in Section 11.4, we can get a power law function if we replace x, y, and z by $\log(x)$, $\log(y)$, and $\log(z)$. We can then solve the equation set in (11.8.2) for A, b, and c. If we use base 10 logs, $a = 10^A$.

The data and their base 10 logs and the necessary log squared are shown in Table 11.1.

TABLE 11.1 Data and their base 10 logs and the necessary squared logs

L	Re	Nu	log10(L)	log10(Re)	(log10(Nu))	(log10(L))^2	(log10(Re))^2	log10(L)* log10(Nu)	log10(Re)* log10(Nu)	log10(L)* log10(Re)
1.81	3250	370	0.2577	3.512	2.568	0.066	12.333	0.662	9.019	0.905
6.40	3330	190	0.8062	3.519	2.279	0.650	12.380	1.837	8.018	2.837
9.47	3700	226	0.9764	3.528	2.354	0.953	12.444	2.298	8.304	3.444
14.1	3500	193	1.1492	3.544	2.286	1.321	12.560	2.627	8.100	4.073
19.4	3670	174	1.2878	3.565	2.241	1.658	12.707	2.885	7.987	4.591
		Sum	4.4773	17.667	11.727	4.649	62.425	10.309	41.428	15.849

This leads to the set of equations for A, b, and c:

5	4.4773	17.667	11.727
4.4773	4.649	15.849	10.309
17.667	15.849	62.4245	41.428

whose solution is $A = -0.700$, $b = -0.343$, and $c = 0.949$.

Since $a = 10^A$, $a = 0.199$. Then the curve fit is

$$Nu = 0.199 * L^{-0.343} * Re^{0.949}.$$

The original N_u and the values resulting from the fit are:

L	Re	Nu	Nu, fit
1.81	3250	370	350
6.40	3300	190	230
9.47	3370	226	205
14.1	3500	193	186
19.4	3670	174	174

The fit is good.

■ 11.9 Splines

Regression attempts to pass a single curve as close as possible to the measured data. The resulting curve provides a convenient way to represent the data and can also provide insight into the physical or chemical processes underlying the measurements.

An alternate approach to regression is to construct a series of lines, each one of which passes through adjacent data. Thus the "fit," called a spline, is perfect, but we achieve this accuracy at the expense of multiple equations. If we have $n + 1$ data pairs, we will have n lines. This process is not useful for analysis, but it is very useful in drawing smooth curves through points in, say, a CAD drawing. It is the mathematical equivalent of expertly drawing a curve through data by hand with a French curve. As in using a French curve, we must pay attention to the intersections of the curve pieces or splines.

To construct a spline we must assume its form. Typically, this is a polynomial. If the polynomial assumed is third degree, we have a cubic spline: for example,

$$S_i = a_i * x^3 + b_i * x^2 + c_i * x + d_i.$$

For each equation we have four unknowns. For n splines we would have $4n$ undetermined coefficients. We have $n + 1$ data pairs that we shall number $i = 0$ to n. Thus

(x_0, y_0) is the first pair and (x_n, y_n) is the last. To obtain the unknown coefficients, we impose the following requirements:

1. The first spline must pass through the first point and the last spline must pass through the last point. This yields two constraint equations.
2. Each spline must pass through the $n - 1$ interior points. This yields $2n - 2$ equations.
3. The first derivatives of intersecting splines must be equal. This yields $n - 1$ equations.
4. At each interior point the second derivatives of intersecting splines must be equal. This yields $n - 1$ more equations.

Adding up the equations from these requirements, we have $2 + 2n - 2 + n - 1 + n - 1 = 4n - 2$ equations.

Since we have $4n$ unknowns, we are two constraints short. For these we may arbitrarily assume that, at x_0 and x_n, the second derivatives are zero. The resulting spline is called a natural spine. Thus, for the second derivative of S_i we have

$$6 * a_1 * x_0 + 2 * b_1 = 0 \text{ at } x_0, \text{ and}$$
$$6 * a_n * x_n + 2 * b_n = 0 \text{ at } x_n.$$

11.9.1 An Abbreviated Spline Example

Suppose we want to pass cubic splines through the first three points of our Cp data. These three points will require two splines, and hence $4 * 2 = 8$ constants and eight equations. Recalling these data from Section 11.3, we have the following:

T	Cp
0	1.008
2	0.999
4	0.998

At $T = 0, Cp = 1.008$, so therefore,

$$a_1 * 0^3 + b_1 * 0^2 + c_1 * 0 + d_1 = 1.008. \tag{11.9.1a}$$

At $T = 2, Cp = 0.999$ through which the first spline must pass. Thus,

$$a_1 * 2^3 + b_1 * 2^2 + c_1 * 2 + d_1 = 0.999. \tag{11.9.1b}$$

The second spline must also pass through this point. Thus,

$$a_2 * 2^3 + b_2 * 2^2 + c_2 * 2 + d_2 = 0.999. \qquad (11.9.1c)$$

At $T = 4, Cp = 0.998$ so

$$a_2 * 4^3 + b_2 * 4^2 + c_2 * 4 + d_2 = 0.998. \qquad (11.9.1d)$$

At $T = 2$, the first derivatives of the two splines must be equal. Thus,

$$3 * a_1 * 2^2 + 2 * b_1 * 2 + c_1 = 3 * a_2 * 2^2 + 2 * b_2 * 2 + c_2, \qquad (11.9.1e)$$

and at $T = 2$ their second derivatives must be equal. Thus,

$$6 * a_1 * 2 + 2 * b_1 = 6 * a_2 * 2 + 2 * b_2. \qquad (11.9.1f)$$

Finally, the second derivatives at $T = 0$ and $T = 4$ are assumed to be 0 (natural spline):

$$6 * a_1 * 0 + 2 * b_1 = 0 \text{ and} \qquad (11.9.1g)$$
$$6 * a_2 * 4 + 2 * b_2 = 0. \qquad (11.9.1h)$$

We have six equations for six unknowns that may be solved to yield

$$S_1 = 0.00075T^3 - 0.0075T + 1.008 \text{ and}$$
$$S_2 = 0.00025T^3 - 0.003T^2 + 0.0105T + 0.998.$$

When these two splines are used in the preceding eight equations, we obtain

at $T = 0, S_1 = 1.008$
at $T = 2, S_1 = 0.999$
at $T = 2, S_2 = 0.999$
at $T = 4, S_2 = 0.998$
at $T = 2$, 1st derivatives, $S_1' = 0.0015; S_2' = 0.0015$
at $T = 2$, 2nd derivatives, $S_1'' = 0.009; S_2'' = 0.009$
at $T = 0$, 2nd derivative, $S_1'' = 0$
at $T = 4$, 2nd derivative, $S_2'' = 0.$

Consequently, our requirements have been satisfied.

■ 11.10 Curve Fitting with Excel

Excel has convenient capability for plotting data and then determining an equation to approximate the data. Suppose we want to establish an equation to represent the specific heat of water, Cp, as a function of its temperature at one atmosphere in the range $T = 0°C$ to $100°C$. We shall use the specific heat of water data from Section 11.5, repeated here for convenience.

T, °C	Cp, kJ/(kg · °C)
0	1.008
20	0.999
40	0.998
60	1.000
80	1.003
100	1.007

These data can be entered into the cells of a spreadsheet. To plot them, we click the "Chart Wizard" icon from the Standard toolbar. Then we must select the type of chart we want. XY (Scatter) lets us plot data columns as a function of the first column of the data. Then the chart subtype lets us plot points, curved lines with or without points, or straight lines, with or without points. The choice of subtype is up to the user. In this example, we selected only points.

After clicking "Next," we must select the range (upper-left corner:lower-right corner) of data to be plotted. We can specify the range by typing it in or by dragging the cursor over the corresponding data in the spreadsheet. Including data titles from the spreadsheet will transfer them to the chart. Following that, we can provide a chart title and chart axis labels, and locate the resulting chart in either the current worksheet or a separate worksheet.

By applying this procedure to the above data in the worksheet, we obtain the data denoted by diamonds and labeled Cp in Figure 11.6.

To find an equation that fits the data, click on the actual Cp data, select the Chart pull-down menu, and then "Add Trendline." This gives us a choice of Linear, Logarithmic, Polynomial, Power, Exponential, and Moving Average curve fits. Choosing "Polynomial" lets us select a polynomial of degree 2 through 6. The "Options" tab lets us display the equation and the correlation coefficient, R^2, (Excel uses R^2 for r^2) on the plot if we choose, which is useful in assessing results. We can also set the y intercept of the curve and specify x or y values for which we want a forecast of the other value using our curve fit.

As always, deciding which curve fit to choose depends on the appearance of the actual data. We can successively choose more than one curve fit for the chart to determine which one does the best job in approximating the given curve. We assess these by the

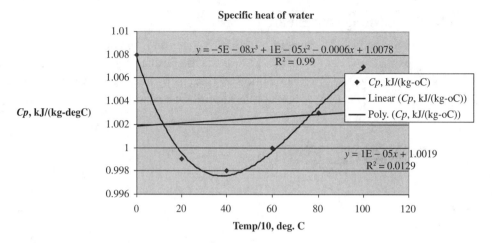

I FIGURE 11.6 Excel Chart with Trendline Showing.

appearance of the fits and comparing their R^2 values. The fit with the highest R^2 value is best.

Applying trendline analysis, for this curve we made both linear and cubic curve fits to the given data. The results, including the equations of the fits and their corresponding correlations are shown in Figure 11.6.

We can see that the linear curve fit does not approximate the data at all and has a correlation coefficient of only 0.0129. The cubic curve fit, on the other hand, approximates the data satisfactorily and has an R^2 value of 0.99, which indicates a very good fit.

■ 11.11 Chapter 11 Exercises

11.1 Thanks to a water treatment chemical, bacteria in a water system have been observed to decline with time as shown. If the normalized concentration with time is given, determine the best curve fit to these data. Consider linear, power, and exponential forms. Begin with a plot of the data. Show all sums.

time	concentration
1	0.5014
3	0.3920
5	0.3065
7	0.2397
9	0.1874
11	0.1466

11.2 The following data were obtained experimentally:

x	y
1	2.5
2	4.25
5	8.38
10	14.21
17	20.86
20	24.0
25	27.55
30	32.0

a. Plot the data. Then select among linear power, and exponential forms.

b. Set up the necessary equations and show all sums.

c. Calculate the coefficients of your fit.

d. Compute the sum of the deviations squared.

e. Calculate the correlation coefficient for these data.

11.3 The following data apply to the specific heat of water:

T	Cp
0	1.008
20	0.999
40	0.998
60	1.000
80	1.003
100	1.007

a. Set up the necessary equations and show sums to make a cubic fit for these data.

b. Calculate the coefficients of your fit.

c. Compute the sum of the deviations squared.

d. Calculate the correlation coefficient for these data.

e. Predict Cp at $T = 75°C$.

11.4 A certain lab experiment produced the following data:

x	y
1	1.45
2	1.41
5	1.37
10	1.18
17	1.09
20	1.01

    a. Plot the data. Then select the curve fit that best represents these data from among linear, power, and exponential forms.

    b. Set up the necessary equations and show all sums.

    c. Calculate the coefficients of your fit.

    d. Compute the sum of the deviations squared.

    e. Calculate the correlation coefficient for these data.

    f. Predict the value of y when $x = 12$.

11.5 In census years, the population of Atlanta, Georgia, (in thousands) since 1950 is as follows:

1950	331
1970	495
1980	425
1990	394
2000	416

    a. Plot the data. Then select the curve fit that best represents these data.

    b. Set up the necessary equations and show all sums.

    c. Calculate the coefficients of your fit.

    d. Compute the sum of the deviations squared.

    e. Calculate the correlation coefficient for these data.

    f. Predict the population of Atlanta in 1978.

11.6 Solve Exercise 11.1 using Excel. Plot the data. Include the coefficients of the fit and the regression coefficient on your plot. Try several fits.

11.7 Solve Exercise 11.2 using Excel. Plot the data. Include the coefficients of the fit and the regression coefficient on your plot. Try several fits.

11.8 Solve Exercise 11.3 using Excel. Plot the data. Include the coefficients of the fit and the regression coefficient on your plot. Try several fits.

11.9 Solve Exercise 11.4 using Excel. Plot the data. Include the coefficients of the fit and the regression coefficient on your plot. Try several fits.

11.10 Solve Exercise 11.5 using Excel. Plot the data. Include the coefficients of the fit and the regression coefficient on your plot. Try several fits.

11.11 Use a cubic spline to fit the first three data pairs of Exercise 11.1.

11.12 Use a cubic spline to fit the last three data pairs of Exercise 11.2.

11.13 Use a cubic spline to fit the first three data pairs of Exercise 11.3.

11.14 Use a cubic spline to fit the last three data pairs of Exercise 11.4.

11.15 Use a cubic spline to fit the first three data pairs of Exercise 11.5.

12 Elliptic Partial Differential Equations

12.1 Introduction

An ordinary differential equation involves only one independent variable. A partial differential equation involves two or more independent variables. For example, the following equation

$$\frac{\partial^2 T}{\partial x^2} + \frac{\partial^2 T}{\partial y^2} = 0 \tag{12.1.1}$$

describes the variation of the temperature, T, in the x and y directions. This equation comes from steady-state conductive heat transfer in a medium with constant properties. The temperature or equivalent must be completely specified on the boundaries. Solving the equation determines the temperature field in the interior of the region. This problem is illustrated in Figure 12.1.

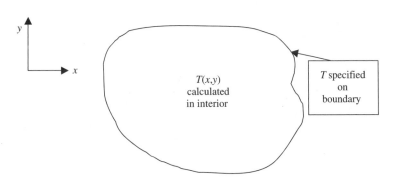

I FIGURE 12.1 Two-Dimensional Boundary Value Problem.

12.2 Derivative Forms

The second derivative is the derivative of the first derivative; that is:

$$\frac{\partial T^2}{\partial x^2} = \frac{\partial\left(\dfrac{\partial T}{\partial x}\right)}{\partial x}.$$

243

Consider a portion of the x-axis that includes the points $r - 1$, r, and $r + 1$. Midway between $r - 1$ and r, and r and $r + 1$, lie $r - \frac{1}{2}$ and $r + \frac{1}{2}$ as shown in the line:

We can express the first derivative of T with respect to x at location $r - \frac{1}{2}, c$ for a given y location (index c) as:

$$\left.\frac{\partial T}{\partial x}\right|_{r-\frac{1}{2},c} = \frac{T_{r,c} - T_{r-1,c}}{\Delta x}. \tag{12.2.1}$$

Since this is a partial derivative, the location of y (denoted by the index c) does not change. Likewise, the derivative at $x = \left(r + \frac{1}{2}, c\right)$ would be

$$\left.\frac{\partial T}{\partial x}\right|_{r+\frac{1}{2},c} = \frac{T_{r+1,c} - T_{r,c}}{\Delta x}. \tag{12.2.2}$$

The second derivative at (r, c) could then be expressed as

$$\left.\frac{\partial^2 T}{\partial x^2}\right|_{r,c} = \frac{\left.\dfrac{\partial T}{\partial x}\right|_{r+\frac{1}{2},c} - \left.\dfrac{\partial T}{\partial x}\right|_{r-\frac{1}{2},c}}{x_{r+\frac{1}{2}} - x_{r-\frac{1}{2}}}. \tag{12.2.3}$$

Observe that $x_{r+1/2} - x_{r-1/2}$ is equal to Δx. Substituting this and equations (12.1.1) and (12.2.1) into (12.2.3), we obtain

$$\left.\frac{\partial^2 T}{\partial x^2}\right|_{r,c} = \frac{\dfrac{T_{r+1,c} - T_{r,c}}{\Delta x} - \dfrac{T_{r,c} - T_{r-1,c}}{\Delta x}}{\Delta x}.$$

Simplifying, this becomes

$$\left.\frac{\partial^2 T}{\partial x^2}\right|_{r,c} = \frac{T_{r+1,c} - 2T_{r,c} + T_{r,c-1}}{\Delta x^2}. \tag{12.2.4}$$

Likewise, for the y direction with fixed x, (r location):

$$\left.\frac{\partial^2 T}{\partial y^2}\right|_{r,c} = \frac{T_{r,c+1} - 2T_{r,c} + T_{r,c-1}}{\Delta y^2}. \tag{12.2.5}$$

Since this equation applies in the interior of the region, with the boundary values defined, it is classified an *elliptic equation*. Another type of partial differential equation, the *parabolic differential equation*, is characterized by one independent variable, such as time, being open-ended. The unsteady heat conduction equation

$$\frac{\delta T}{\delta t} = \alpha \frac{\delta^2 T}{\delta x^2}$$

is a parabolic equation. We shall not consider such equations.

This chapter provides only a brief introduction to the numerical solution of partial differential equations, which is a book-length topic in itself. The classic treatment by Smith (1965) is a good place to continue your study.

■ 12.3 An Elliptic Partial Differential Equation Example

Applying the results from Equation (12.1.1) dealing with temperature variation, we obtain

$$\frac{\partial^2 T}{\partial x^2} + \frac{\partial^2 T}{\partial y^2} = \frac{T_{r+1,c} - 2T_{r,c} + T_{r-1,c}}{\Delta x^2} + \frac{T_{r,c+1} - 2T_{r,c} + T_{r,c-1}}{\Delta y^2} = 0.$$

If we choose a square grid—that is, $\Delta x = \Delta y$—and simplify, we obtain

$$T_{r+1,c} + T_{r-1,c} + T_{r,c+1} + T_{r,c-1} - 4T_{r,c} = 0 \tag{12.3.1}$$

or

$$T_{r,c} = \frac{T_{r+1,c} + T_{r-1,c} + T_{r,c+1} + T_{r,c-1}}{4}. \tag{12.3.2}$$

Thus, on a square grid, the value of $T_{r,c}$ is equal to the average of its four neighbors.

Since $T_{r,c}$ is influenced by each of its neighbors, as they change it will change. Thus the values are connected (recall the Gaussian–Seidel method in Chapter 9), and we must solve Equation (12.3.2) over the x-y grid by iteration.

In a given iteration using VBA, we can sweep over the x-y grid using nested For loops. But when do we stop iterating? For each iteration, we compute the change in T at each location ($\Delta T_{r,c}$). We then find the largest of these changes in absolute value for that iteration and continue to iterate until the largest change is less than a previously chosen tolerance.

■ 12.4 Elliptical Equation Solver Flowchart

A flowchart for this procedure is shown here in Figure 12.2.

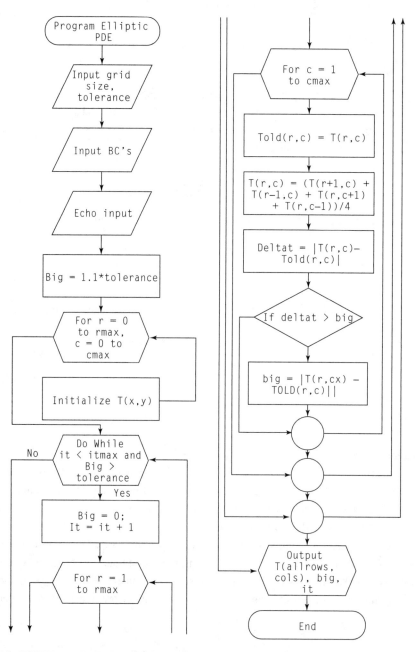

I FIGURE 12.2 Elliptic Partial Differential Equation Solver Flowchart.

■ 12.5 Solving Elliptic Differential Equations with Excel

Excel can be used to solve a set of elliptic differential equations of the form of Equation (12.3.2), repeated here for convenience:

$$T_{r,c} = \frac{T_{r+1,c} + T_{r-1,c} + T_{r,c+1} + T_{r,c-1}}{4}.$$

We again recognize that each $T_{r,c}$ is the arithmetic average of its four neighbors. For a temperature stored, say, in Excel cell D6, this means its value is

$$= (C6 + E6 + D5 + D7)/4. \tag{12.5.1}$$

The cell arrangement is as follows:

```
              D5
      C6      D6      E6
              D7
```

Similar expressions apply for all the other values in the interior of the region. Fixed boundary conditions apply on the exterior of the computation boundary surrounding the interior. Because the interior points are mutually dependent, changing one potentially will change many values, which necessitates iteration. Fortunately, Excel has the built-in capability to iterate automatically on such a set.

We begin by defining each cell value to be calculated in terms of its neighbors and prescribing the boundary values that surround them. Suppose we define a region with $T = 100$ along the top, $T = 75$ along the left side, $T = 50$ along the right side, and $T = 0$ along the bottom, with 10×10 (i.e., 100) interior cells. The field would resemble Table 12.1, where equations of the form of (12.3.2) would replace the ?'s.

Note that the corner values are missing. These could be defined in terms of either their row or their column, creating a potential conflict. But Equation (12.3.2) does not use values diagonal to the center cell, hence the corner values are not needed.

When these values are entered, Excel will generate a warning about circular references. This is because cells are mutually dependent—for example, D6 depends on E6, and E6 depends on D6. Thus the references are circular, but we can ignore the warning and eliminate the problem by iterating on the interior cells.

Let us continue by highlighting the interior cells. Then, from the pull-down Tools menu, select "Options/Calculations." In the gray dialog box that displays (see Figure 12.3), select "Automatic", "Iteration," the "Maximum iterations" (default 100), and the "Maximum change" (default 0.001.) When "Calc Now" is selected or "F9" is pressed, Excel will repeat-

Table 12.1 Excel Solution Setup

	100	100	100	100	100	100	100	100	100	100	
75	?	?	?	?	?	?	?	?	?	?	50
75	?	?	?	?	?	?	?	?	?	?	50
75	?	?	?	?	?	?	?	?	?	?	50
75	?	?	?	?	?	?	?	?	?	?	50
75	?	?	?	?	?	?	?	?	?	?	50
75	?	?	?	?	?	?	?	?	?	?	50
75	?	?	?	?	?	?	?	?	?	?	50
75	?	?	?	?	?	?	?	?	?	?	50
75	?	?	?	?	?	?	?	?	?	?	50
75	?	?	?	?	?	?	?	?	?	?	50
	0	0	0	0	0	0	0	0	0	0	

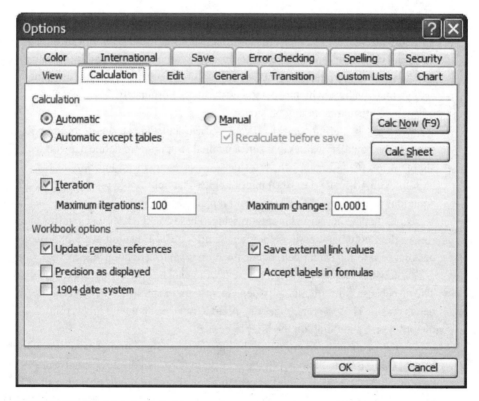

I **FIGURE 12.3** Excel Calculation Options.

edly recalculate the highlighted range of the interior. Calculations will cease when either the maximum number of iterations has been performed or the largest change within those cells, from one iteration to the next, is less than the specified tolerance.

After applying the iterative calculation with Excel, we obtain the results shown in Table 12.2.

Table 12.2 Excel Solution

	100	100	100	100	100	100	100	100	100	100	
75	86.36	90.18	91.44	91.67	91.41	90.77	89.64	87.61	83.55	74.31	50
75	80.25	82.94	83.89	83.84	83.18	82.02	80.19	77.24	72.27	63.71	50
75	76.69	77.46	77.34	76.62	75.47	73.93	71.85	68.91	64.57	58.27	50
75	74.05	72.85	71.40	69.82	68.15	66.38	64.39	61.96	58.82	54.79	50
75	71.66	68.48	65.61	63.10	60.94	59.06	57.37	55.72	53.98	52.07	50
75	69.10	63.82	59.45	56.02	53.44	51.56	50.29	49.56	49.32	49.51	50
75	65.93	58.24	52.35	48.11	45.23	43.46	42.68	42.90	44.21	46.67	50
75	61.37	50.86	43.59	38.84	35.90	34.36	34.07	35.16	37.97	42.95	50
75	53.70	40.23	32.33	27.75	25.18	24.02	24.10	25.68	29.56	37.16	50
75	38.19	24.04	17.75	14.64	13.06	12.42	12.61	13.92	17.41	26.14	50
	0	0	0	0	0	0	0	0	0	0	

■ 12.6 Derivative Boundary Conditions

In the previous discussion and examples, we have specified the value of the function on the surrounding boundary and obtained its interior values by approximately solving the partial differential equation. But it is possible that the derivative of the function, rather than its value, will be specified on the boundary. Mathematicians call boundary-specified values the *Dirichlet condition*; derivative boundary conditions they call *Neumann boundary conditions*.

Assume that at some boundary point (r,c) the normal derivative of T is specified. This may be in either the x or the y direction. For example, assume that it is in the y direction and its value is a. Then we have

$$\left.\frac{\partial T}{\partial y}\right|_{x} = a.$$

If we approximate this derivative with a first central difference, we will have

$$\left.\frac{\partial T}{\partial y}\right|_{x} = \frac{T_{r,c+1} - T_{r,c-1}}{2\Delta y} = a$$

where $T_{r,c+1}$ would be above the boundary.

Solving for $T_{r,c+1}$, we obtain $T_{r,c+1} = 2\Delta y * a + T_{r,c-1}$.

Recalling Equation (12.3.1),

$$T_{r+1,c} + T_{r-1,c} + T_{r,c+1} + T_{r,c-1} - 4T_{r,c} = 0,$$

and substituting for $T_{r,c+1}$ from the boundary condition, we obtain

$$T_{r+1,c} + T_{r-1,c} + 2\Delta y * a + T_{r,c-1} + T_{r,c-1} - 4T_{r,c} = 0.$$

Simplifying, we have

$$T_{r+1,c} + T_{r-1,c} + 2\Delta y * a + 2 * T_{r,c-1} - 4T_{r,c} = 0,$$

$$T_{r,c} = \frac{T_{r+1,c} + T_{r-1,c} + 2\Delta y * a + 2 * T_{r,c-1}}{4}.$$

It is this equation that we would have to solve on the y boundary for $T_{r,c}$.

The corresponding equation for a normal derivative equal to b along the right x-boundary would be

$$T_{r,c} = \frac{2 * T_{r-1,c} + T_{r,c+1} + T_{r,c-1} + 2\Delta x * b}{4}.$$

Similarly, along $y = 0$

$$T_{r,c} = \frac{T_{r+1,c} + T_{r-1,c} - 2\Delta y * a + 2 * T_{r,c+1}}{4},$$

and along $x = 0$

$$T_{r,c} = \frac{2 * T_{r+1,c} + T_{r,c+1} + T_{r,c-1} - 2\Delta x * b}{4}.$$

If a point at (r,c) is an upper-right corner with derivative boundary conditions in both the x and y directions, we obtain

$$T_{r,c} = \frac{2 * T_{r-1,c} + 2 * T_{r,c-1} + 2\Delta x * b + 2\Delta y * a}{4}.$$

Of particular interest is a symmetry boundary condition. In this case, the normal derivative at the boundary is zero, hence a or $b = 0$.

If a temperature region were symmetric in either the x or the y direction, we could solve it by considering only one half of the field and applying the symmetry boundary condition along the axes of symmetry. If it were symmetric in both the x and y directions, we could apply the symmetry boundary condition along two axes and consider only one quarter of the field.

■ 12.7 Chapter 12 Exercises

12.1 Consider the region and boundary conditions shown in Figure 12.4. Divide the region into 12×12 interior cells with $\Delta x = \Delta y$ and solve Equation (12.3.1) using a program following the flowchart in Section 12.4.

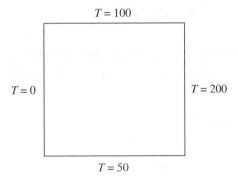

I FIGURE 12.4 Geometry of Problems 12.1–12.3.

12.2 Solve Exercise 12.1 using 18×18 interior cells and a computer program following the flowchart in Section 12.4.

12.3 Solve Exercise 12.1 using the relaxation capability in Excel.

12.4 Solve Exercise 12.1 with the boundary condition shown in Figure 12.5.

12.5 The boundary conditions given in Figure 12.5 are symmetric. Consider only one-quarter of the entire region and solve it, applying zero-derivative boundary conditions along the lower and right-hand sides of your reduced region using a computer program following the flowchart in Section 12.4.

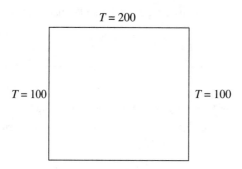

I FIGURE 12.5 Geometry of Problems 12.4–12.5.

12.6 Consider the region and boundary conditions shown in Figure 12.6. Divide the region into 12×12 interior cells with $\Delta x = \Delta y$ and solve Equation (12.3.1) using your program.

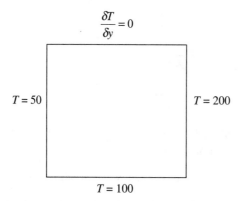

I FIGURE 12.6 Geometry of Problems 12.6–12.7.

12.7 Solve Exercise 12.5 using 18×18 interior cells and your computer program.

12.8 Solve Exercise 12.5 using the relaxation capability in Excel.

Excel Basics

Microsoft Excel is a commercial spreadsheet. Spreadsheets allow you to manipulate data, arranged in cells like squares on a big checkerboard, in a variety of ways. Calculations can be made by referring to the data by their cell locations and operating on them using formulas, either your own or formulas supplied by Excel. Many features allow you to copy data from one range of cells to another, to plot graphs, and to perform financial or statistical analyses. Spreadsheets can even be used for math such as numerical integration and solving differential equations, although there are more efficient ways to do this.

Excel lets you select many screen options with the mouse. The options are displayed as small pictures or icons. Excel also offers documentation, available in the program through "Help" commands. The quickest way to find help on a given topic is by clicking Help at the top right of the screen and then choosing "Contents and Index." Type in the keyword of your topic of interest, then select among the variations offered.

Names of an icon in the Standard toolbar—the row below the File, Edit, and other menus at the top of the window—will display if the mouse is dragged over it and kept there briefly. To read a description of the icon's function, select Help and type in the icon name.

1. To activate Excel: (1) Click the "Start" icon at the lower left of the screen (or depress the "Start" button—the Windows icon between the Ctrl and Alt keys to the left of the space bar or between the Alt and the Edit keys to the right of the space bar); and (2) after the menu displays, select "Microsoft Office Excel" (or "All Programs," then "Microsoft Office/Microsoft Office Excel"). Alternatively, click the "Excel shortcut" icon if it appears on the desktop.

   The Excel worksheet consists of menus at the top and a portion of the cell grid that can be used. Letters identify cell columns; numbers identify cell rows.

2. Data and text can be entered directly into cells as desired. Use the mouse to move to the cell you want and click the left button. The active cell will be surrounded by a black outline. Alternatively, you can move directly to a particular cell by typing in the cell location in the white space above A—for example, C10—and pressing "Enter."

Data, text, and formula results, are placed to the right side of the cell (right justified; and text is left justified) unless you choose otherwise. To center the data, select the icon in the second row, two to the right of the U (Underline) icon and click it. This icon shows a small page in which the lines are centered. The icon to its left aligns data at the left of the cells, the icon to its right aligns data at the right of the cells.

3. **Cell widths** in a specific column or columns can be changed to accommodate more information by (1) clicking a cell in the column(s), (2) clicking "Format" in the toolbar above the icons, and (c) selecting "Column" then "Width" in the pull-down menu. Type in the desired number of spaces in the "Column Width" dialog box and click OK. Cell widths can also be changed by dragging the border of the region containing the letter at the top of a column.

   Cell gridlines can be turned on or off by selecting "Tools," then "Options," and then clicking on the gridlines box to toggle it on or off.

4. **Arithmetic formulas** can be entered into cells to use data in other cells. Formulas *must* start with an equal (=) sign. For example, suppose cell C5 contains the value 11 and C6 contains the value 15. If you want the sum of these to be displayed in cell C7, (1) select "C7" with the mouse, (2) type =C5+C6, (3) and press "Enter."

   In formulas, the + sign is used for addition, − for subtraction, * for multiplication, and / for division. Exponentiation is marked with a caret ($^\wedge$, over the 6). For example, to indicate "x squared," type x$^\wedge$2. Parentheses should be used to group operations. For example, the sum of A and B divided by the sum of C and D would be =(A+B)/(C+D). If you typed A+B/C+D, your result would be A and D added to B and divided by C.

5. An important feature of spreadsheets is their ability to **reference cells** either absolutely or relatively in formulas. Absolute cell coordinates (row, column, or both) are preceded by a dollar sign ($). For example, $C5 means that the C row reference is fixed or absolute, C$5 means the 5 column reference is absolute, and $C$5 means both row and column references are absolute. If a cell reference is not absolute (no $), it is relative. This means that wherever the cell reference is used, data from cells in the same relative location as in the defining cell will be used.

   Suppose you want to add together the numbers in cells A1 and B1 and put the result in C1; likewise, with A2 and B2 into C2, A3 and B3 into C3, and so on. In cell C1 you might type =A1+B1 to produce the result in C1. Now you could copy this formula into cells C2, C3, and so forth using the "Copy" command, and the results would be computed and placed in the appropriate cells. Because you used relative addressing (no $'s), C2 would contain the results of adding together the two cells immediately to its left, just as was done with cell C1. Thus Excel would maintain the relative cell positions when the formula is copied.

   On the other hand, if you wanted to add the results of A1 to B1 with the results in C1, A1 to B2 with the results in C2, A1 to B3 with the results in C3, and so on, you could type $=\$A\$1+B1$ in cell C1. Then when you copied this formula into the other cells, the value in A1 would always be used. In other words, you used the absolute location, A1, in the formulas rather than the corresponding relative locations, but the relative location of the B cells.

6. Material in a cell or range of cells may be **copied** to another portion of the spreadsheet. (1) Highlight the cells to be copied, (2) select the "Edit" menu, then "Copy" from the toolbar, (3) highlight the cells into which you want the material to be copied, and (4), press "Enter." The number of cells copied must equal the number of cells into which the data are to be copied. Otherwise, an error message will display stating that size and shape must be maintained.

   This procedure will copy a *formula* or text into the cells selected. To **copy values**, begin in the same way but, after selecting the target cells, click the "Edit" menu and then "Paste Special." From the menu that displays, select "Values" and click "OK."

7. Because summing of spreadsheet data is so common, the Σ icon (AutoSum) is available on the first icon row (the Standard toolbar). To use it, select the cell where you want the sum to display, click the Σ icon, and enter the range to be summed—for example, A1:A26 to sum the total in column A, rows 1 through 26. You may also drag the mouse over the data to be summed and include a cell for the result.

   Access other **Excel formulas** by selecting the "Function Wizard." This is the icon with the f_x just to the right of the Σ in the Standard toolbar. Click this icon. The left part of the menu lists formula categories and the right side lists formulas in each category in alphabetical order. The categories are Financial, Math & Trig, Engineering, Most Recently Used, and All. (Bessel functions are included in the Engineering list.) Data needed can be entered in a command line in the cell or by filling in the dialog box that displays for the function selected. A brief description of the formula is displayed below the left-hand list. Click the formula you want and select "Next." Filling in the menu that displays will provide the data needed to evaluate the formula.

   The dialog box that displays allows you to search for a function by category or to select a function directly from the complete list.

   Note that the square root function in Excel is **sqrt()** (this differs from VBA where it is sqr()). Also, pi is **PI()**. The base of the natural log system, e, is **exp()**. The natural log is **ln()** and the common log is **log10()**. The number operated on goes into the parentheses. (Nothing is needed for pi.) A number such as 0.0056 may be entered in scientific notation as 0.56e-2.

8. **Data may be sorted** by highlighting the cells (more than one column is possible) in question and then selecting "Sort" from the Data pull-down menu. Sort lets you specify the column to be sorted first, and then subsequent columns to distinguish among equal values in the first column sorted. If a range of columns is selected, data

in the column sorted first will be the most important in determining the order of all the columns. If only a single column is to be sorted, you can highlight the column and click either the "Sort Ascending" (A Z ↓) or the "Sort Descending" (Z A ↓) button in the Standard toolbar.

9. You can define a **reference table** in a portion of your spreadsheet and then use values in it elsewhere. Type =VLOOKUP(A,B,C,D) in the cell location where you want the looked-up value to display. Here, A is the independent value to be compared to the values in the first column of the reference table, B contains the range of the table, C is the column in the table whose dependent value you want to use, and D is TRUE or FALSE. If it is TRUE, the largest value not exceeding A in the table will be used. If FALSE, only exact matches will be considered.

Example: =VLOOKUP(6.5,$A$6:$B$12,3,TRUE) means select, from the third column in the table defined in cells $A$6 through $B$12, the value corresponding to the largest value not exceeding 6.5 in the first column of that table. NOTE: Data in the table *must* be in ascending order.

10. To put a **time/date stamp** on your program in a cell, use NOW( ). Cell A1 is convenient for this. Put your name in the same region. If the time format is abnormal (e.g., format displays as a 5-digit number), (1) select the "Format" pull-down menu, (2) then "Cells," (3) then choose Time or Date in the "Category" list on the left, and (d) choose the desired format in the "Type" list on the right.

11. **Charts (graphs)** can be made with the "Chart Wizard." Select this option by clicking its icon in the Standard toolbar and place the chart where you want it on the spreadsheet by clicking there. You can highlight the data fields with the mouse before invoking the Chart Wizard or with the second interactive box (labeled "Step 2 of 4") after invoking it. The first dialog box inside the Wizard lets you select the chart type. Standard *x-y* plots are called *scatter diagrams* by the Chart Wizard. If this is what you want, click "XY" (Scatter) in the list on the left side and then select the desired data form (lines, points, lines and points, and so on) on the right. Click "Next" and move to the data screen. You can select the data range if you haven't done so previously, by entering the range now (*x* values, then *y* values in that order). Click "Next" and add titles, axes, labels, and so on, as desired. Click "Next" to move to the final Wizard box. This allows you to place the chart in the current worksheet or on a worksheet by itself.

12. To **print** your spreadsheet, select the "File" pull-down menu and "Print." A dialog box will display giving you choices on which sheets you want printed, how many copies, and so on. Select these as needed and then click the "OK" button.

You can select the orientation of your spreadsheet as horizontal (landscape) or vertical (portrait) by clicking the "File" menu and selecting "Page Setup." You then specify your choice by clicking on the appropriate "bullet."

Print **cell formulas** instead of numbers by selecting "Tools/Options/View/and Formulas." This will display the formulas in their respective cells. This form of the spreadsheet can then be printed as just described.

13. To **save** a document, select the "File" pull-down menu, then click "Save As." Next to the "Save in" window, click the scroll down arrow and then the drive where you want to save the file. To the right of the "File name" window, type in the name that you want to assign to the file. The default file type is .xls. To save as this type requires no additional work. If you want to change to another type, such as a WK4 (Lotus 1-2-3) file, click the scroll down arrow opposite "Save as type" and select the desired file type. Then click "Save."

    To **retrieve** a document, do the following. Select the "File" menu, then click "Open." Next to the "Look in" window, click the scroll down arrow and then the appropriate drive and/or folder. Select the appropriate file from the list that appears to the right of "File name" Click "Open."

14. To **leave** Excel after saving any material you want saved, click "Exit" in the File menu.

Many more details about Excel are available in Krishan, S. I. *Computing with Excel and VBA: A Problem Solving Approach*, 2nd Edition. Jones and Bartlett, 2009.

B Computer Representation of Numbers

Humans have 10 fingers on their two hands. Thus, it is natural that our dominant number system is based on 10. (Others do exist. The Sumerians, for example, used a system based on 60!) But what does the decimal system really mean? Consider the number 25: 25 means $2 * 10^1 + 5 * 10^0 = 20 + 5$.

Similarly, $125 = 1 * 10^2 + 2 * 10^1 + 5 * 10^0$. If we had a maximum of five slots in which to write a decimal number we might have the following representation:

10^4	10^3	10^2	10^1	10^0

0	0	1	2	5

We have unused slots in the 10^4 and 10^3 locations.

The biggest number we could write in this format would be

9	9	9	9	9

What about fractions? Places to the right of the decimal place can be expressed as:

10^{-1}	10^{-2}	10^{-3}	10^{-4}	10^{-5}

Thus, $0.125 = 1 * 10^{-1} + 2 * 10^{-2} + 5 * 10^{-3} = 0.1 + 2 * 0.01 + 5 * 0.001$
$= 0.1 + 0.02 + 0.005$.

This could be stored in our slots as

1	2	5	0	0

In contrast to numbers greater than zero, unused locations are at the right end of the slots.

Computer logical units typically are on/off devices. Therefore, numbers in computers are generally based on a binary system, that is, zero and one. Numbers in binary are strings of nothing but zero's and one's. For example, 25 is 11001. We humans do not easily read these.

How is this decimal equivalence established? $11001 = 1 * 2^4 + 1 * 2^3 + 0 * 2^2 + 0 * 2^1 + 1 * 2^0 = 16 + 8 + 0 + 0 + 1 = 25$. Thus, 25 could be stored in a computer as the 5-bit character number

1	1	0	0	1

If the computer word is 8 bits long, with the leftmost bit reserved for the sign of the number, 0 for a positive number and 1 for a negative, $+25$ would be

0	0	0	1	1	0	0	1

We would store $-127 = -(1 * 2^6 + 1 * 2^5 + 1 * 2^4 + 1 * 2^3 + 1 * 2^2 + 1 * 2^1 + 1 * 2^0)$ as

1	1	1	1	1	1	1	1

This would be the most negative number that could be written in an 8-bit machine without using scientific notation. For a 16-bit machine, however, the most negative number would be -32768.

For numbers larger or algebraically smaller than the storage bit limits permit, the computer could use floating-point arithmetic: that is, numbers of the form

$$m * b^{ee}$$

where m is the mantissa, b the base, and ee the exponent. The mantissa is written such that its leftmost (leading digit) is not 0. In scientific notation, the base is 10 so we would write 0.0125 as $0.125 * 10^{-1}$ and 125 as $0.125 * 10^3$.

To fully represent a floating-point number, we would have to provide information on its sign, the sign of its exponent, the exponent, and the mantissa. This information might be arranged as

sign	exponent sign	exponent	mantissa

If we use 0 for $+$ and 1 for minus, $0.125 * 10^3$ base 10 would be as follows:

0	0	3	125

The number 10^3 in binary is 1111101000 and 0.125 is .001.

Thus, 123 would be written as

0	0	1111101000	001

We see that even relatively small numbers require a lot of space to write in binary. Thus computer designers limit the size of the numbers that can be written even in floating-point form. We saw in Chapter 2 that the largest floating-point number that can be written as a single-precision floating-point number is 3.402823×10^{38}.

A problem can occur even if we are working with numbers much smaller than the storage limits for the language. Using base 10, because it is more familiar and numbers can be written much more compactly, suppose we want to add 987 to 124, both written in scientific notation:

$$0.124 \times 10^3$$
$$\underline{0.987 \times 10^3}$$
$$1.111 \times 10^3$$

The exponent is no different from that of the input numbers, but we now have four digits in our sum. If our computer is restricted to three digits, the 1.111 would be stored as 1.11 and the rightmost digit would be lost. This is familiar to us as round-off. Even carrying many places does not prevent us from losing information.

If the rightmost digit had been five or more, the computer might either have dropped it entirely (called *chopping*) or the one to its left would have been rounded up to a two. Either way accuracy would be lost. Chopping is cheaper to accommodate in a computer. Excel performs a rounding function in cell displays as VBA does in `MsgBox`.

C | VBA Command Summary

■ C.1. Assignment Statements (Chapter 2)

Arithmetic: addition (+), subtraction (−), (multiplication (*), division (/), exponentiation (^).

Assignment statements store the result of the calculation right of the equal sign in the location of the variable on the left—that is, $c = a + b$. Operations are performed left to right. The hierarchy is exponentiation, multiplication/division, and addition/subtraction. Parentheses should be used to group calculations. (See Chapter 2.)

■ C.2 Input/Output (Chapter 2)

Input from `InputBox`: `a = Val(InputBox("Enter a now."))`
    Input from Excel cell A3: `a = cells(3,1).value` or
                                     `a = cells(3,"a").value`
    Input from file (after first opening the file)

```
Open "f:\problem1inp.txt" for input as #4
Input #4, a
```

    Output (labeled) with `MsgBox`: `MsgBox "a = " & a`
    Output to Excel cell B4: `cells(4,2).value = a` or
                             `cells(4,"b").value = a`
    Output to a file (after opening file #7)

```
Open "f:\problem1out.txt" for output as #7
Print #7, a or
Write #7, a
```

    Close files when no longer using.

    Print statements may be formatted. #'s indicate the number of places left and right of the decimal place.

```
Print #7, format(Pi, "#.####") would display a previously defined
value of Pi as 3.1416.
```

■ C.3 Loops and Decisions (Chapter 4)

With a known number of repetitions, n is the increment (step value); omit if 1 (n may be <0).

```
For i = first to last step n
    ....
Next i
```

Conditional

```
Do While (condition is true)
    ....
Loop
```

Boolean symbols: $>$, $<$, $>=$, $<=$, $<>$, and, or

```
If (condition is true) then
        ....
Else
    ....
Endif
```

■ C.4 Subprograms and Functions (Chapter 6)

In calling sub: `Call sub2(argument list)`

```
Sub sub2(argument list)
    ....
End Sub
```

In calling sub: `y = f1(argument list)`

```
Function f1(argument list)
    ....
    f1 = ....
End function
```

Main VBA functions: `sin, cos, tan, atn, sqr, abs, exp` for e^x, `log(x)` for $\ln(x)$

```
Pi = 4*atn(1) for π
```

Excel functions in VBA programs (see Chapter 6): example of pi.

```
Application.Worksheetfunction.functionname()
example Application.Worksheetfunction.pi()
```

■ C.5 Arrays (Chapter 10)

```
Dim AA(10,11), as double, X(10) as double
```

AA is a two-dimensional array, X a one-dimensional array; both are declared as double-precision data types.

■ C.6 Miscellaneous

1. The security level can be changed using Tools/Macro/Security. Selecting "Medium" allows you to respond to the dialog box each time you open a file. If you change the security level, it will not apply until the *next* time you open Excel.

2. To access VBA, select "Tools/Macro/Macro." (It is better *not* to select "VBA Editor." Otherwise, you might have trouble accessing the VBA file in the future.) Then create a new VBA file or edit an existing one. At the bottom of the dialog box that displays, you can highlight a particular macro if more than one exists and select "This Workbook" (generally preferred) or "All Open Workbooks."

3. Put file opens near the top of the program (but after the initial Sub name()). Close files after you are finished with them but not inside a loop.

4. Save input and output files as type .txt.

5. To automatically clear a specified spreadsheet range at the beginning of execution in VBA, use the following where the actual range to be cleared should replace a1:z100:
   ```
   Range("a1:z100").clearcontents
   ```

6. Use Control Break (together) to halt execution if stuck in a loop.

7. More than one command can be put on one line if they are separated with the following, for example:
   ```
   a = 6: b = 7: c = 9
   ```

8. Scientific notation can be entered using e for exponential, such as 0.5e-06. The base e of the natural log system is exp as
   ```
   x = exp(-6)
   ```

9. To control the format in MsgBox, use the following line of code where the number of #'s to the left of the decimal point control the number of integers and the number of #'s to the right control the number of decimal places.

Msgbox format (label variable, "##.#####"); for example,

MsgBox Format ("a = " & a & " b = " & b,"##.######"),

would label and write a and b each with up to two figures left of the decimal and six after.

10. Debug Pull-down Menu

Select "Run to cursor." The program stops at that line *without* executing it. Therefore, you should select the line below the one of interest, even if you have to create a do-nothing line like "JWH = 1." Values generated up to that point can be read by dragging, and holding, the mouse over them.

D Glossary

(Absolute) uncertainty: Δu or du the uncertainty calculated either exactly (Δu) or approximately (du); it is absolute as opposed to relative and hence has the units of u.

Approximate error: $E_a = (x\text{new} - x\text{old})$ compares change in successive values during iteration; it has the units of x.

Approximate relative error: $\varepsilon_a, (x\text{new} - x\text{old})/x\text{new} * 100$ computes change in successive values during iteration and compares this to the latest value; it is a ratio and hence does not have units.

Approximate uncertainty analysis: Calculate du, the derivative of the quantity whose uncertainty is desired, in terms of the derivatives of its factors.

Bound the error: Calculate the remainder term in a Taylor series and choose the value of $z(x_0 <= z <= x)$ such that it maximizes the magnitude of the derivative used. This is the maximum error in the Taylor series approximation.

Diagonal dominance: Diagonal coefficients are greater than, or equal to, the sum of the magnitudes of the other coefficients in the row. At least one row is needed where the diagonal coefficient is greater (sufficient, but not necessary, condition for Gaussian–Seidel convergence).

Exact uncertainty analysis: $\Delta u = (u\text{max} - u\text{min})/2$.

Expand about: x_0, the point in a Taylor series expansion where everything is known.

Maclaurin series: A Taylor series expanded about $x_0 = 0$.

Normalization: In Gaussian–Jordan, divide the equation through by the diagonal coefficient; in Gaussian–Seidel, divide the equation through by the largest coefficient in the row left of the equal sign; in Gaussian elimination, DON'T normalize.

Pivot: In Gaussian elimination and Gaussian–Jordan, this involves interchanging two rows so that the largest coefficient (in magnitude) is on the main diagonal for the column being considered; only two rows are exchanged; a pivot is done only as a column is being considered and not before.

Rearrangement: In Gaussian–Seidel, this involves rearranging equations to try to get the largest coefficient in each row left of the equal sign on the main diagonal; rearrangement might involve moving more than two equations.

Relative uncertainty: The uncertainty is calculated either exactly or approximately and then divided by the mean value: for example, $\dfrac{\Delta u}{\bar{u}} * 100$; it has no units.

Tolerance: $0.5 * 10^{(2-sf)}$ where sf is the number of significant figures.

True error: E_T, the difference between the correct value and the approximation; generally it is not known since the correct value is not known.

Numerical Methods with the Casio fx-115MS Calculator

This calculator is approved for use on the Fundamentals of Engineering (FE) examination.

■ E.1 Solving a Nonlinear Equation

This calculator can be used to solve a single nonlinear equation of the user's choice. It requires entering the equation and a starting value.

To illustrate, consider the example equation from Chapter 7,

$$f(x) = \tan(x) * \sin(x) - 2.15 = 0.$$

Key in the equation as:

tan

Alpha x

X

Sin

Alpha x

−

2.15

Shift Solve

1.47   (This is the starting value.)

=

Shift Solve

The labeled answer, $x = 1.167$, displays on the screen shortly.

■ E.2 Solving a Quadratic or Cubic Equation

This calculator will solve quadratic and cubic equations and yield both real and complex solutions.

Equations are written in the form

$$ax^3 + bx^2 + cx + d = 0 \text{ or}$$
$$ax^2 + bx + c = 0.$$

Press the Mode kcy until EQN displays. Press 1. When "Unknowns" displays, press the cursor key on the right side. When degree appears press 3 (2 if a quadratic). The calculator then prompts for the values of a, b, c, and d.

Consider the cubic equation

$$x^3 + 2x^2 + 3x - 4 = 0.$$

On the calculator:

a? press 1 =
b? press 2 =
c? press 3 =
d? press (−) 4 =

Immediately, x1 = .7760 displays. Pressing the down cursor yields x2 = −1.388. The Re R ⇔ I at the top right of the screen indicates that this root is complex. Pressing Shift Re ⇔ Im (over the = key) yields the complex part, 1.797i. Pressing the down cursor again yields x3 = −1.388, and Shift Re <−> yields −1.797i.

The variables x2 and x3 have this form because complex roots always have ± complex parts.

■ E.3. Solving Linear Equations

This calculator will solve sets of up to three linear equations of the form

$$a1 * x + b1 * y + c1 * z = d1,$$
$$a2 * x + b2 * y + c2 * z = d2, \text{ and}$$
$$a3 * x + b3 * y + c3 * z = d3.$$

Press the Mode key until EQN appears. Press 1. When "Unknowns" displays, press 3 (2 if 2 linear equations). The calculator then prompts for the values of a1, b1, c1, d1, and so on. Enter each coefficient value followed by =.

For the set

$$0x_1 - 2x_2 + 5x_3 = 12,$$
$$x_1 - 6x_2 + 2x_3 = 9, \text{ and}$$
$$4x_1 - 1x_2 + 1x_3 = 4,$$

after the value of 4 is entered, the calculator displays x = 0.2871. Pressing the down cursor yields y = −0.7525 and, again, z = 2.099.

■ E.4 Numerical Integration

This calculator will also perform numerical integration. It requires entering the integrand, the upper and lower limits, and N, where 2^N intervals are used. For the example as presented in Chapter 5,

$$\int_1^5 x^2 \ln x\, dx,$$

begin by pressing the $\int dx$ key. Then enter the integrand, a, b, and the number of intervals, separated by commas as follows.

$\int dx$
Alpha x
x^2
X
Ln x

,

1

,

5

,

6

)

=

The parentheses ()) ahead of the equal sign (=) balances the parentheses (() automatically inserted by the calculator after $\int dx$ was entered. This calculation will take several moments but then the answer, 53.282, displays.

Excel Functions in VBA

A
Acos
Acosh
And
Asin
Asinh
Atan2
Atanh
AveDev
Average

B
BetaDist
BetaInv
BinomDist

C
Ceiling
ChiDist
ChiInv
ChiTest
Choose
Clean
Combin
Confidence
Correl
Cosh
Count
CountA
CountBlank
CountIf
Covar
CritBinom

D
DAverage
Days360
Db
DCount
DCountA
Ddb
Degrees
DevSq
DGet
DMax
DMin
Dollar
DProduct
DStDev
DStDevP
DSum
DVar
DVarP

E
Even
ExponDist

F
Fact
FDist
Find
FindB
FInv
Fisher
FisherInv
Fixed

Floor
Forecast
Frequency
Ftest
Fy

G
GammaDist
GammaInv
GammaLn
GeoMean
Growth

H
HarMean
HLookup
HypGeomDist

I
Index
Intercept
Ipmt
Irr
IsErr
IsError
IsLogical
IsNA
IsNonText
IsNumber
Ispmt
IsText

K
Kurt

L
Large
LinEst
Ln
Log
Log10
LogEst
LogInv
LogNormDist
Lookup

M
Match
Max
MDeterm
Median
Min
MInverse
MIrr
MMult
Mode

N
NegBinomDist
NormDist
NormInv
NormSDist
NormSInv
NPer
Npv

O
Odd
Or

P
Pearson
Percentile

PercentRank
Permut
Phonetic
Pi
Pmt
Poisson
Power
Ppmt
Prob
Product
Proper
Pv

Q
Quartile

R
Radians
Rank
Rate
Replace
ReplaceB
Rept
Roman
Round
RoundDown
RoundUp
RSq
RTD

S
Search
SearchB
Sinh
Skew
Sln
Slope
Small
Standardize
StDev

StDevP
StEyx
Substitute
Subtotal
Sum
SumIf
SumProduct
SumSq
SumX2MY2
SumX2PY2
SumXMY2
Syd

T
Tanh
Tdist
Text
Tiny
Transpose
Trend
Trim
TrimMenu
TTest

U
USDollar

V
Var
VarP
Vdb
VLoopup

W
Weekday
Weibull

Z
Ztest

G | Differentiation Fundamentals

At a minimum, differentiation is needed in defining a series and assessing the error in numerical differentiation. Two rules need to be remembered to do this successfully:

The product rule $\dfrac{d(f(x) * g(x))}{dx} = f * \dfrac{dg}{dx} + \dfrac{df}{dx} * g.$

Example: $h(x) = \sin(x) * \cos(x),$

$$\dfrac{dh}{dx} = \sin(x) * \dfrac{d\cos(x)}{dx} + \dfrac{d\sin(x)}{dx} * \cos(x)$$

$$= \sin(x) * (-\sin(x)) + \cos(x) * \cos(x), \text{ and}$$

$$\dfrac{dh}{dx} = \cos^2(x) - \sin^2(x).$$

The chain rule $\dfrac{df(g(x))}{dx} = \dfrac{df}{dg}\dfrac{dg}{dx}.$

Example: $f(x) = \tan^3(x) = g(x)^3$ where $g(x) = \tan(x).$

$$\dfrac{df}{dx} = 3 * g(x)^2 * \dfrac{dg}{dx}$$

$$= 3 * \tan^2(x) * \dfrac{d\tan(x)}{dx}, \text{ and}$$

$$\dfrac{df}{dx} = 3 * \tan^2(x) * \sec^2(x).$$

Frequently, we have to apply both rules.

Example: $f(x) = \tan^3(x) * \sec(x).$

$$\dfrac{df}{dx} = \dfrac{d\tan^3(x)}{dx} * \sec(x) + \tan^3(x) * \dfrac{d\sec(x)}{dx}$$

$$= 3 * \tan^2(x) * \dfrac{d\tan(x)}{dx} * \sec(x) + \tan^3(x) * \dfrac{d\sec(x)}{dx}$$

$$= 3 * \tan^2(x) * \sec^2(x) * \sec(x) + \tan^3(x) * \sec(x) * \tan(x), \text{ and}$$

$$\dfrac{df}{dx} = 3 * \tan^2(x) * \sec^3(x) + \tan^4(x) * \sec(x).$$

Cash–Karp Runge–Kutta VBA Program for Two Ordinary Differential Equations

```
Option Explicit
Sub C_Kode()
'     *** FILENAME CKode ***
'     LAST MODIFIED 1/17/08
'     TO SOLVE ORDINARY DIFFERENTIAL EQUATIONS USING
'     THE CASH-KARP (RUNGE-KUTTA-FEHLBERG) METHOD. INPUT
'     ARE INITIAL AND FINAL INDEPENDENT VARIABLES, INITIAL
'     DEPENDENT VARIABLES, STEP-SIZE, PRINT-FREQUENCY, AND
'     ERROR TOLERANCE.
'     THESE ARE IN THE SPREADSHEET. OUTPUT IS TO THE
'     SPREADSHEET AND A FILE.

'     X IS THE INDEPENDENT VARIABLE; Y AND Z ARE THE
'     DEPENDENT VARIABLES.
'     THE DERIVATIVES ARE DEFINED IN FUNCTIONS F (DY/DX)
'     AND G (DZ/DX).
'     THE ERROR IS CHECKED DURING PRINT-OUT. IF THE ERROR
'     IS AN ORDER OF MAGNITUDE DIFFERENT FROM THE TOLERANCE
'     THE STEP-SIZE IS ADJUSTED.

      Dim FF(6) As Double:   Dim GG(6, 6) As Double
      Dim XKY(6) As Double:  Dim XKZ(6) As Double
      Dim T(4)As Double:     Dim TT(5) As Double

      Dim x As Double:     Dim y As Double:     Dim z As Double
      Dim dx As Double:    Dim x1 As Double:    Dim tol As Double
      Dim pf As Double:    Dim pp As Double:    Dim y4 As Double
      Dim z4 As Double:    Dim xold As Double:  Dim yold As Double
      Dim zold As Double:  Dim eay As Double:   Dim eaz As Double
      Dim i As Integer:    Dim ic As Integer:   Dim im1 As Integer
      Dim j As Integer:    Dim pd As Integer

      Open "RKFOUT.TXT" For Output As #7
      Range("A4:H100").ClearContents
      tol = 0.00001
      x = Cells(2, 1).Value
```

```
y = Cells(2, 2).Value
z = Cells(2, 3).Value
x1 = Cells(2, 6).Value
pf = Cells(2, 7).Value
dx = Cells(2, 8).Value
tol = Cells(2, 9).Value
Print #7,"X0 =", x," Y0 =", y," Z0 =", z
Print #7,"XL =", x1," DX =", dx," PF =", pf," tol =", tol
Print #7," X"," Y"," Z",
Print #7," EAY"," EAZ"," DX"

'    FF'S ARE THE STEP FRACTIONS
FF(1) = 0#
FF(2) = 0.2
FF(3) = 0.3
FF(4) = 0.6
FF(5) = 1#
FF(6) = 7 / 8


'    GG'S ARE THE K WEIGHTINGS
GG(2, 1) = 0.2
GG(3, 1) = 3# / 40#
GG(3, 2) = 9# / 40#
GG(4, 1) = 0.3
GG(4, 2) = -0.9
GG(4, 3) = 1.2
GG(5, 1) = -11 / 54
GG(5, 2) = 2.5
GG(5, 3) = -70 / 27
GG(5, 4) = 35 / 27
GG(6, 1) = 1631 / 55296
GG(6, 2) = 175 / 512
GG(6, 3) = 575 / 13824
GG(6, 4) = 44275 / 110592
GG(6, 5) = 253 / 4096


'    T'S ARE THE WEIGHTINGS IN Y4
T(1) = 37 / 378
T(2) = 250 / 621
T(3) = 125 / 594
T(4) = 512 / 1771


'    TT'S ARE THE WEIGHTINGS IN Y5
TT(1) = 2825 / 27648
TT(2) = 18575 / 48384
TT(3) = 13525 / 55296
```

```
TT(4) = 277 / 14336
TT(5) = 0.25

ic = 4
pp = pf: pf = pf + x
pf = pf + x
pd = 1
Do While Abs(x - x1) > dx / 10 And x < x1
    xold = x
    yold = y
    zold = z
    For i = 1 To 6
        x = xold + dx * FF(i)
        y = yold
        z = zold
        If i > 1 Then
        im1 = i - 1
            For j = 1 To im1
                    y = y + dx * GG(i, j) * XKY(j)
                    z = z + dx * GG(i, j) * XKZ(j)
                Next j
            End If
        XKY(i) = F(x, y, z)
        XKZ(i) = G(x, y, z)

        If (pd = 1) Then
                Cells(2, 9 + i).Value = XKY(i)
                Cells(3, 9 + i).Value = XKZ(i)
        End If
    Next i
    y4 = yold + dx * (T(1) * XKY(1) + T(2) * XKY(3) + T(3) *
            XKY(4) + T(4) * XKY(6))
    z4 = zold + dx * (T(1) * XKZ(1) + T(2) * XKZ(3) + T(3) *
            XKZ(4) + T(4) * XKZ(6))
    y = yold + dx * (TT(1) * XKY(1) + TT(2) * XKY(3) + TT(3)
            * XKY(4) + TT(4) * XKY(5) + TT(5) * XKY(6))
    z = zold + dx * (TT(1) * XKZ(1) + TT(2) * XKZ(3) + TT(3)
            * XKZ(4) + TT(4) * XKZ(5) + TT(5) * XKZ(6))
    eay = 1 - y4 / y
    eaz = 1 - z4 / z
    x = xold + dx
    pd = pd + 1
    If (Abs(pf - x) < dx / 10# Or x > x1) Then
            Cells(ic,"A").Value = x
            Cells(ic,"B").Value = y
            Cells(ic,"C").Value = z
```

```
                    Cells(ic,"F").Value = eay
                    Cells(ic,"G").Value = eaz
                    Cells(ic,"H").Value = dx
                    Print #7, x, y, z, eay, eaz, dx
                    If Abs(eay) < tol / 10 And Abs(eaz) < tol /
                        10 Then
                            dx = 2 * dx
                            pp = 2 * pp
                    End If
                    If Abs(eay) > 10 * tol And Abs(eaz) > 10 *
                    tol Then
                            dx = dx / 2
                            pp = pp / 2
                    End If
                    pf = pf + pp
                    ic = ic + 1
                End If
          Loop
          Close #7
End Sub
Function F(x, y, z)
     F = 8.8 - 0.00137 * y ^ 2
End Function
Function G(x, y, z)
     G = y
End Function
```

Bibliography

Cash, J. R. and A. H. Karp, *ACM Transactions on Mathematical Software*, 16:201–222, 1990.

Chapra, S. C. and R. P. Canale, *Numerical Methods for Engineers*, 5th ed., McGraw-Hill, 2006.

Colebrook, C. F., "Turbulent Flow in Pipes, with Particular Reference to the Transition between the Smooth and Rough Pipe Laws," *J. of the Institute of Civil Engineers*, London, 1939.

Dorn, W. S. and D. D. McCracken, *Numerical Methods with Fortran IV Case Studies*, Wiley, 1972.

Gear, C. W. "The Automatic Integration of Stiff Ordinary Differential Equations," *Proc. IFIP Congress*, supplement, booklet A, 1968.

Haaland, S. E. "Simple and Explicit Formulas for the Friction Factor in Turbulent Pipe Flow," *J. of Fluids Engineering*, March 1983, pp. 89–90.

Hiestand, J. W. and A. R. George, "Generalized Steady State Method for Stiff Equations," *AIAA Journal*, vol. 14, no. 9, 1976, pp. 1153, 1154.

Jaluria, Y. and K. E. Torrance, *Computational Heat Transfer*, Hemisphere, 1986.

Krishan, S. I. *Computing with Excel and VBA: A Problem Solving Approach*, 2nd Edition. Jones and Bartlett, 2009.

Lomax, P., *VB and VBA in a Nutshell*, O'Reilly, 1998.

McCracken, D. D. and W. S, Dorn, *Numerical Methods and Fortran Programming*, Wiley, 1961.

Mueller, J. P., *VBA for Dummies*, 4th ed. Wiley, 2003.

Press, W. H., B. P. Flannery, S. A. Teukolsky, and W. T. Vetterling, *Numerical Recipes*, Cambridge Univ. Press, 1989.

Scarborough, J. B., *Numerical Mathematical Analysis*, 6th ed., Johns Hopkins Univ. Press, 1966.

Scheid, F., *Numerical Analysis*, 2nd ed., Schaum's Outline, McGraw-Hill, 1989.

Smith, G. D., *Numerical Solution of Partial Differential Equations with Exercises and Worked Solutions*, Oxford Univ. Press, 1965.

Walkenkenbach, J., *Excel Power Programming with VBA*, 2nd ed., IDG Books, 1996.

INDEX

Gauss Elimination

$$2x_1 + 8x_2 + 2x_3 = 22$$
$$x_1 + 5x_2 + 9x_3 = 26$$
$$-10x_1 + 3x_2 + 7x_3 = 17$$

Pivot R1 & R3

$$\begin{bmatrix} -10 & 3 & 7 & | & 17 \\ 1 & 5 & 9 & | & 26 \\ 2 & 8 & 2 & | & 22 \end{bmatrix}$$

$$M = \frac{a_{21}}{a_{11}} = \frac{1}{-10}$$

$$-\tfrac{1}{10} R_1 = [1 \quad -0.3 \quad -0.7 | -1.7]$$

$$M = \frac{a_{31}}{a_{11}} = \frac{2}{-10} = -\tfrac{1}{5}$$

$$-\tfrac{1}{5} R_1 = [2 \quad -0.6 \quad -1.4 \ | -3.4]$$

$$\begin{bmatrix} -10 & 3 & 7 & | & 17 \\ 0 & 5.3 & 3.4 & | & 25.4 \\ 0 & 8.6 & 9.7 & | & 27.7 \end{bmatrix}$$

Pivot R3 & R2

$$\begin{bmatrix} -10 & 3 & 7 & | & 17 \\ 0 & 8.6 & 3.4 & | & 25.4 \\ 0 & 5.3 & 9.7 & | & 27.7 \end{bmatrix}$$

$$M = \frac{a_{32}}{a_{22}} = \frac{5.3}{8.6} = 0.616$$

$$0.616\,(R_2) = 0 \quad 5.298 \quad 2.094 \quad 15.646$$

$$\begin{bmatrix} -10 & 3 & 7 & | & 17 \\ 0 & 8.6 & 3.4 & | & 25.4 \\ 0 & 0 & 7.606 & | & 12.054 \end{bmatrix}$$

$$X_3 = \frac{12.054}{7.606} = 1.585$$

$$X_2 = \frac{[25.4 - (3.4 \cdot 1.585)]}{8.6} = 2.327$$

$$(-1)^{\#\,of\,pivots} (diagonals) = det$$

$$(-1)^2 (-10)(8.6)(7.606) = -654.116$$

its minimizes round off error and avoids division by 0.

by the 5th iteration there is virtually no change in answers through 3rd sig. fig. if accurate, 12 iterations required.

$$X_1 = \frac{[17 - (2.327 \cdot 7) - (1.585 \cdot 3)]}{-10} = 0.404$$